'Retrieving Wenty's Sturty Bird'

THE STORY OF BREAM CENOTAPH 1921 - 2001

To all who served;
those who didn't come home,
those who did
and those who stood and waited

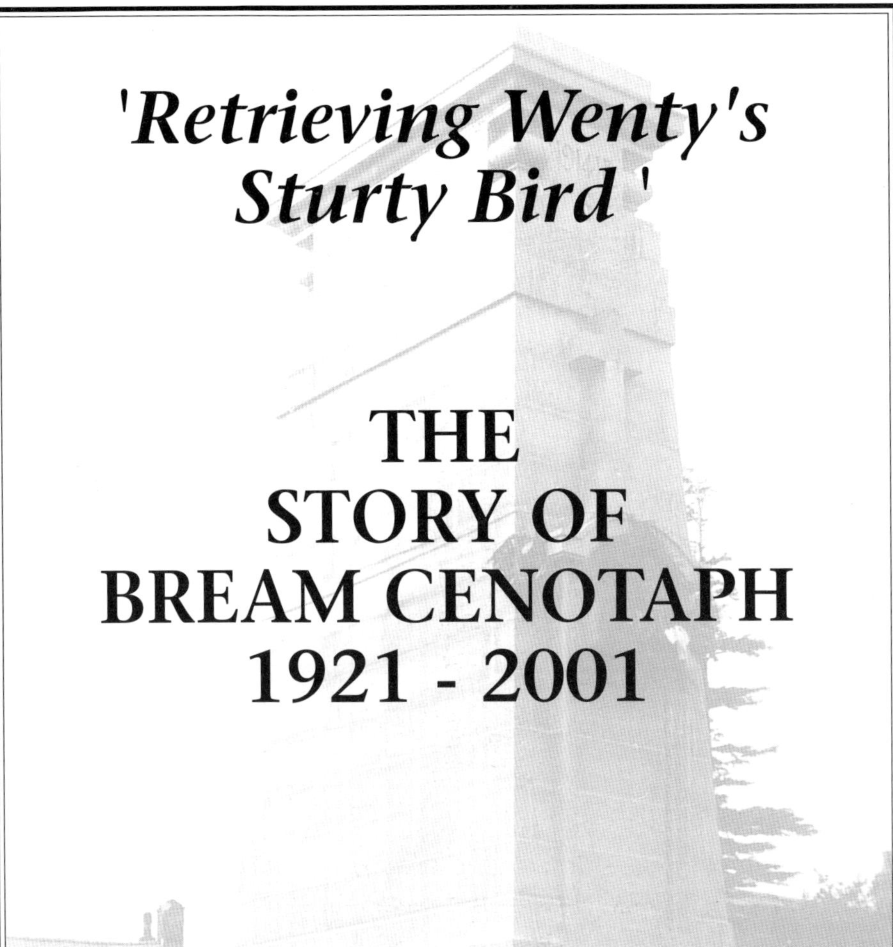

'Retrieving Wenty's Sturty Bird '

THE STORY OF BREAM CENOTAPH 1921 - 2001

Ian Hendy

Black Dwarf Publications

The cenotaph shortly after completion. As yet there are no names.
Photo courtesy Melville Watts

CONTENTS

Designed by Neil Parkhouse. Cover design by Ade Haines, Artytype

British Library Cataloguing-in-Publication Data. A catalogue record for this book is available from the British Library
ISBN 1 903599 03 2

Black Dwarf Publications
47 – 49 High Street, Lydney, Gloucestershire GL15 5DD

Printed by M.D. Jenkins Ltd.,
Unit 53/54, Lydney Trading Estate, Harbour Road, Lydney, Gloucestershire GL15 4EJ

Leslie Lodge, wounded at Ypres, 9th October 1917, died Bradford War Hospital, 26th January 1918. His sister, Rose, particularly remembered his beautiful dark brown eyes – "*like velvet*". *Photo courtesy Nancy Merry*

Introduction

The idea for this book started when a flyer came through the letterbox last summer, outlining the Cenotaph Restoration Project and asking for fundraising ideas. As an erstwhile History Teacher with a particular interest in World War One, I thought a book about the Cenotaph and the names on it would be a useful way of raising money to help restore the monument. That original thought still holds good and every penny made by the sale of this book will go into the Restoration Fund.

There are other reasons for writing but, before explaining them, I should point out what the book is not. It is not an attempt to glorify war. War may be portrayed, usually by politicians, as 'necessary', or 'just', or even as 'inevitable' but in human terms it is invariably a tragedy. Lives, usually young lives, are lost; people are maimed and scarred, emotionally as well as physically; children grow up without parents, parents grow old without children. Those who believe war is glorious should visit the Thiepval Memorial to the Missing of The Somme, containing over 73,000 names of men with no known grave. Or they should go to the Menin Gate in Ypres, with over 54,000 similar names. A further 35,000 names of the missing at Ypres are recorded on the wall of the cemetery at Tyne Cot. We do well to remember these appalling statistics. As George Santayana wrote, '*Those who cannot remember the past are condemned to repeat it.*'

Rudyard Kipling, who lost his only son in World War One, was asked to write the inscription for the Stone of Remembrance that would be situated in the larger cemeteries. He chose '*Their name liveth for Evermore*' taken from the *Book of Ecclesiasticus* in *The Apocrypha*. This is the principle reason for writing the book, to perpetuate the names of those who died. And 'names' not just as labels as in a telephone directory but names in the wider sense of 'reputation'. Bream Cenotaph is not a list of the dead. It is a list of the people who died. There is a difference. Every one of those names was a person and their story should be told.

Alice Davies said to me when I was talking to her about this book, "*When we are gone our story will die with us"*. My aim is to ensure that that does not happen. When future generations of children from this district look at Bream Cenotaph and (hopefully curious!) ask "*Why is that there?*" or "*What are those names?*", if this book provides some answers, albeit imperfect and incomplete ones, I shall be satisfied.

As a historian I believe individuals cannot be studied in isolation from their environment. I have therefore tried to take individuals' stories against the backcloth of the village at the time and both against the wider context of the two World Wars that dominated those times.

Obviously there have been constraints of time and money in producing this book. The restoration of the Cenotaph will not wait forever till I finish writing. Similarly, the costs of research have been met by the limitations of a personal budget, thus not as much time has been spent in the Public Record Office as I would have ideally liked.

Nevertheless, within constraints, everything of a factual nature has been checked. I accept that there may be sentences that will produce a "*Thou bist wrong*" or a "*Chun right*" response. If any error is proved I apologise unreservedly. Similarly if anything written causes offence. Such was never my intention.

In researching this book I came across a reference in the Log Book of Bream Mixed School, dated October 25 1915, to Balaclava Day. The battle of Balaclava took place during the Crimean War (1854-6) against Russia and is now best remembered for the Charge of the Light Brigade. In 1915, Bream School was given a holiday on Balaclava Day for fundraising. Every boy was encouraged to work and donate his day's earnings to help provide a new Ambulance Wagon for the war. [(1)]

That Balaclava Day Log entry reminded me of the wider issue of remembrance. How many of us today know of Balaclava or commemorate it as they did in 1915? And yet the time distance between 1915 and Balaclava is similar to the distance between us and the early years of the Second World War. How will future generations view our commemoration of Remembrance Sunday; with the same detachment – or ignorance – as we ourselves view Balaclava Day?

The answer to this question is the key to remembrance. If we record the stories behind the names before they are forgotten, then, even though first hand memory will fade, the people will live on because their story has been told. Story telling brings a form of immortality.

I hope this book will appeal to everyone with an interest in Bream. Wartime did produce many tragedies but it also produced a lot of humour and some situations that could be either tragic or humorous or both! In this category could be included some village scenes of the inter-war years, like Downham's Bakery delivery vans being loaded by Frank and Albert Downham – Frank with one arm and Albert with one leg. Similarly, in the inter-war years many villagers attended Mullans the one-eyed dentist. All were wartime injuries and no doubt borne with courage and then – as now – a little teasing!

There are some situations where the humour is timeless. Like the new Forest recruit who was asked by his officer "*A grenade has just landed by you, what steps would you take?*". Back came the answer "*Big buggers in the opposite direction*". (2)

It really has been a privilege to research and write this book. I have spoken to a lot of people, sometimes about painful memories and they have been nothing but helpful and supportive. I will never forget the magic of listening to Basil James recount the story of how, on 5th August 1914, he went with his father, Dapper, in a pony and trap to Coleford to watch the Territorials parade out of town.

I have been fortunate not only in having so much support but also that individuals and families in the community have memories that go back a long way. The village network was – and is! – formidable and could usually find me answers to the most obscure questions. Amazingly some people I visited even knew the questions I was going to ask before I asked them, probably even before I had thought of them! Such is the remarkable village network and many thanks to it. I'm convinced that had Hitler's troops occupied Bream in 1940 they would not have stood a chance against 'the network'!

My sincere thanks to all who have helped. In a very real sense this is their book. I hope I have acknowledged them all in the references, if I haven't my sincere apologies for my oversight. Specifically, I would like to thank the staffs of Gloucestershire Record Office and Caldicot, Cinderford and Ross Libraries. My particular thanks go to the staff of *The Forester* for their co-operation and support in the use of back copies of Forest newspapers, to Andrew Ferris for his work on the photographs and to Pam Smith for deciphering a manuscript that would, at times, have frustrated the enigma code-breakers! Thanks, too, to my colleagues on the Cenotaph Restoration Committee for all their encouragement. I consider myself lucky to have been part of the project. Also my thanks to the four points of my compass, Alice Cook, Hylton Miles, Dave Rudge and Geoff Wildin who, when I've been stuck, have always been happy to point me in the right direction.

My final thanks to my daughters, Karen, Louise, Ella and Laura for their constant encouragement and practical assistance, and my long suffering wife Catherine, who, having given birth four times, knows something of the agony and the ecstasy a project such as this entails.

Ian Hendy
Bream
April 2001

A Note on Money

Currency in Britain during the two World Wars was pre-decimal and expressed in pounds, shillings and pence, i.e. 12 pence (12d) to the shilling (1/- or 1s) and 20 shillings (240 pence) to the pound. Throughout this book I have used pre-decimal currency and, for the sake of clarity, included the decimal equivalent in brackets.

Obviously, the purchasing power of the pound has changed considerably since 1914. £100 in 1914 = £250 in 1920 = £160 in 1938 = £260 in 1946 and = £5,800 in 1998.(1)

In 1914 the average income was less than £2 per week and the average family rent was 5/- (25p) per week.(2) By 1938 the average male manual worker earned £3.9s.0d (£3.45) in an average 47.7 hour week. By 1945 this had changed to £6.1s.4d (£6.07) for an average working week of 49.7 hours. Prices doubled between 1938 and 1946.(3)

Publisher's Note

We are now over eighty years from the Armistice which ended the First World War and fifty-five years from the end of the Second. The author has done a tremendous job in tracing the men and women on the Cenotaph Roll of Honour and, where possible, has located photographs as well. Readers will appreciate, however, that many of these pictures have not stood the passage of time that well, whilst some only exist as images published in newspapers of the time. As a result, many of the pictures within these pages are of poor quality but it was deemed important to use them nevertheless. Such images, by their grainy and faded nature, pass their own silent comment on those times and on those who did not live to grow old.

Wenty's Sturty Bird

Having used 'Wenty's Sturty Bird' in the title of this book it is interesting to speculate how many in the village can work out the meaning of it. I suspect a lot more recognise the name 'Wenty' than know what a 'Sturty Bird' is.

Many individuals are named after famous people and this was how Wenty came to be Wenty. He was christened Wentworth Sidney Cooper, the Wentworth coming from Charles Wentworth Dilke, the MP for the Forest and the man after whom the Dilke Hospital is named. Interestingly, in trawling through the 1891 Census for the Forest, I came across Charles Dilke lodging at Speech House. In answer to 'occupation' he replied *'Privy Councillor'* – rather a rare census entry!

Wenty was born in 1910 and lived all his life in the village until his death in December last, aged 90. He was very highly regarded by all who knew him and was a mine of information when it came to local history. Wenty was suggested to me as someone I should talk to about the Cenotaph and I was very fortunate to spend a couple of afternoons with him a few weeks before he died. It really was a privilege to talk to him about the village.

It was Wenty who told me about the Sturty Bird and I was intrigued. The game – for that is what Sturty Bird is – has virtually died out in the last fifty years but it is a very ancient game, being first found in Egypt circa 2500 BC. It is played under a variety of rules and is known by a variety of names. In Britain its more usual name is 'Tipcat' or 'Nipsy' (or 'Bat and Catty' in Monmouthshire). To my present knowledge I have found no reference to the game being called 'Sturty Bird' in any other area. The name may even be localised to this part of the Forest or even this village. (1)

As described to me by Wenty the game was played thus. The 'bird' (or 'cat' elsewhere) was a piece of wood some 4 inches to 6 inches long, about as thick as one's thumb and pointed at both ends. The bird was placed on the ground, it was then hit on the end with a stick and as it flew in the air, hopefully within reach, it was then hit again. The hitter then estimated how many stick lengths the bird had flown, say 20 sticks. If your opponent challenged your estimate by saying *"Mint em"* (measure them), you then used the stick to check the distance. If your estimate was correct to a stick's length the points went on your score. Then it was your opponent's turn.

Sturty Bird was not a game for the playground nor for any place where the 'bird' or 'cat' could damage property or hurt passers-by. It was a game for open spaces.

Sun Tump was the ideal location for playing Sturty Bird. Wenty Cooper was playing the game there in 1921. Of course Sun Tump was not as ideal a location for the game as it had been the year before. Now, in 1921, a cenotaph was being built but no harm could come to anybody in playing the game if the builders were not there. Inevitably, Wenty's Sturty Bird flew into the hollow cenotaph, which was so high by that stage that he could not retrieve it.

And so Wenty's Sturty Bird became part of the fabric of the Cenotaph from that day to this. When the Cenotaph is dismantled, if Wenty's Sturty Bird hasn't rotted, it should be possible to recover it. I am only sorry that Wenty will not be around to see it and retrieve his Sturty Bird, eighty years after its disappearance.

Wenty Cooper on his 90th birthday. *Photo courtesy Angus Cooper*

THE FIRST WORLD WAR

The First World War Cenotaph Roll of Honour

GREATER LOVE HATH NO MAN THAN THIS,
THAT A MAN LAY DOWN HIS LIFE FOR HIS FRIEND

A. BROOKES
A. DODGSHON
E. EVANS
P. FROWEN
H.L. HOWELLS
T. JONES
W.T. KEAR
J. KENT
E. KILBY
L. LODGE
E. LUCAS
L. MAYNE
J. MATTY
G. NELMES
W.H. PEARCE
A.E. THOMAS
H. WASLEY
A. WATKINS
O. WILKS
E. WORGAN
G. WYNN

The Maypole, Bream, photographed shortly before the outbreak of the First World War. *Photo courtesy The Archive Shop*

1914

"Over By Christmas . . ."

The assassination of the heir to the Austrian throne, Archduke Franz Ferdinand, and his wife, Sophie, on a visit to Sarajevo in Bosnia on 28th June 1914, was to precipitate a chain of events that would lead to the outbreak of war in less than six weeks. Austria, supported by her ally, Germany, blamed Serbia for the murder. Russia, supported by her ally France, gave support to Serbia. By the 1st of August, Austria, Germany, Russia and France had all mobilised and Germany had begun to implement The Schlieffen Plan, which was quickly to involve Britain in the war.

The Schlieffen Plan had been drawn up in 1905 to cater for a situation in which Germany would be involved in a war on two fronts – a Western Front against France and an Eastern Front against Russia. The Plan assumed that Russia would be slow to mobilize and that Germany should aim to knockout France within six weeks before confronting Russia. This knockout blow against France would be best achieved by invading France through neutral Belgium. Thus Germany invaded Belgium on August 4th.

Britain had been a signatory of the 1839 Treaty of London, which guaranteed the independence and neutrality of Belgium and, alarmed by the prospect of a potentially hostile Germany in control of the entire opposing channel coastline, issued them with an ultimatum to leave Belgium. That ultimatum expired at midnight on August 4th and Britain was thus involved in its first major war (with the possible exception of the Crimean War of 1854-6) since the end of the Napoleonic Wars in 1815.

Within a few days of the start of the war, the B.E.F. (British Expeditionary Force), or 'Old Contemptibles' as they preferred to be called, had crossed the Channel and 70,000 Reservists had been called to the Colours. War Minister Kitchener called for a new army of volunteers which produced a massive response – 500,000 had come forward by mid-September.

The B.E.F., after an initially successful delaying action at Mons (23-24 August), was forced to fall back to avoid isolation. Early in September the

German advance on Paris was halted at the Battle of the Marne. The Schlieffen plan had failed. Both sides then tried unsuccessfully to outflank each other and by mid-November a line of trenches extended 450 miles from the Channel coast to Switzerland. The Western Front had come into being, and was to claim hundreds of thousands of lives over the next four years.

At the time of these momentous events the population of Bream was 2,524 (1911 Census) and the principal occupation was mining.(1) There is no doubt that everyone would have been aware of what was happening on the Continent and a headline in *The Dean Forest Mercury* as early as 7th August 1914 refers to 'The Great War'.

The 'Bream Items' in *The Mercury*, dated 14th August, gives a real insight into the village at that time:

> *'The war seems to have made little difference in the village. Everything seems to be going on as before. The Rugby and Association Football Clubs are making arrangements for the coming season, and practice matches will soon commence. The tradespeople of Bream are showing a splendid spirit, and although there seemed an inclination on the part of a few people to rush the various supply stores, the check was immediately put on, and the customers were at once informed that nobody, whether rich or poor, would be allowed any more than their usual weekly allowance. Prices have remained normal except in a few instances. Special prayers were offered in all the Nonconformist Churches on Sunday, and there were special hymns and prayers at the service in connection with St. James's parish church in the evening. A united prayer meeting was held at the United Methodist Church on Tuesday evening by the three Nonconformist Churches, when a good number was present. The Rev. J.E. Squires conducted the service and gave a short, appropriate address. There will be a special service at St. James's Church on Thursday evening. Several young men have enlisted during the week out of the village of Bream, and no fewer than 97 men living in the village have had the call to go and serve their country. The Bream Amateurs' A.F.C. will feel the loss of their clever left-half, Mr. G. Cox, who has volunteered for his country.'*

The same newspaper informed its readers that the Severn Railway Bridge was now under guard. The following week, *The Mercury*, as part of Kitchener's campaign, was advertising 'Your King and Country Need You', appealing for any man between 19 and 30 to volunteer (by 11th September the age limit had been raised to 35). Recruiting Stations were established at Cinderford, Coleford and Lydney. On Friday, 11th September, *The Mercury* reported 105 new recruits at Coleford in the previous week and over 200 who had enlisted at Lydney Police Station since early August. The first to volunteer from Bream was E. Ashton, from Mill Hill, on 8th August.

On 12th September, a meeting was convened at the Rising Sun by Mr. J.J. Dodgshon of Bream Lodge to set up a Miniature Rifle Club in Bream. The Club was officially opened on Saturday, 31st October, the landlord Mr. A.H. Morse lending the clubroom free of charge. It was open for practice, 10-12am, 7-9pm and Thursday afternoon. They had two rifles and two targets.(2)

On 9th October, *The Mercury* stated that 52 men had now left the Princess Royal colliery to serve their country. On 20th November, the paper named 30 volunteers from Bream who had signed up.

By the end of November the realities of war were beginning to have a human impact on the village. A Belgian Refugee Fund was set up and there were already two wounded Belgian soldiers in the village.(3) On 16th December, the village welcomed Mr. and Mrs. Auguste de Coninch as refugees.(4) Word had also come through that Fred Meek of Bream Rugby Club and Mr. Alfred Brooks of Little Drybrook had both been wounded on the Western Front.(5) Clearly the war would not be "*over by Christmas*" – as popularly believed in August – and yet, as far as the village was concerned, 1914 had been a good year. There had, after all, been no fatalities.

The Territorials Farewell Parade in Coleford on Wednesday 5th August 1914. Joe Kent, later to die in 1918 in a German p.o.w. camp, came to watch these men leave (see page 61). *Photo courtesy The Archive Shop*

'B' Company, 13th (Forest of Dean) Battalion, Gloucestershire Regiment, practising trench digging at Malvern in 1915. *Courtesy The Archive Shop*

1915

"Only 4 Million Germans Left to Fight"

The year 1915 saw a growing deadlock on the Western Front. The war was becoming one of attrition to be won by the side with the greater reserves of manpower. A good example of this was a lecture delivered at The Shaftsbury Theatre, in London, by Colonel F.M. Manders. Manders argued that the Germans could put eight-and-a-half million men into the field. To date their losses were 4,200,000 and therefore there were not many more than four million to hold the two fronts. The lecture was reported in *The Mercury*, 2nd July, under the headline '*Only 4 Million Germans left . . .*'

On the Western Front whatever attacks were made, e.g. at Neuve Chappelle in March-May 1915 or Loos in September 1915, resulted in few gains and many casualties. In April, the German Army came near to achieving a breakthrough in the second battle of Ypres following the use of a new and deadly weapon – poison gas. The war also came to the Home Front as air raids started when initially Zeppelin airships and later bombers attacked British cities. From this point onwards, civilians were potential victims of war even though they were many miles from the front line.

The year also saw a number of attempts to influence the outcome of the war through 'sideshows', initiatives well away from the Western Front which would make a decisive breakthrough. These occurred in Salonika, Mesopotamia and Italy, but the most notable sideshow was Gallipoli. Turkey had joined the war on the side of Germany and Austria in November 1914. By the spring of 1915 Britain and France's ally, Russia, was flagging. It was felt that a successful attack on Turkey would relieve pressure on Russia and possibly on the Western Front if Turkey had to be reinforced by Germany. A force of British, Australian and New Zealand troops was landed on the Gallipoli peninsula in Turkey in April. Further landings were made in August but, after suffering over 200,000 casualties, the troops were withdrawn in December.

There were two fatalities from the Parish of Bream in 1915, Alfred Brookes and Leslie Main, the former on the Western Front, the latter in Gallipoli.

Both men are commemorated on two monuments, on Bream Cenotaph and in the Theatre of War where they met their deaths. Remarkably, in both cases the surname is spelt differently on both monuments, 'Brookes' on Bream Cenotaph is 'Brooks' on Le Touret Memorial, France. Leslie Main's commemoration is further complicated by a different initial as well as surname. Thus the L. Mayne of Bream Cenotaph is commemorated on the Helles Memorial in Turkey as J. Main. They are one and the same person.

The spelling of Brooks with an 'e' or without has caused local confusion for years. Alfred's own father's birth certificate in 1869 is in the name of Brookes, yet his death certificate in 1934 calls him Brooks. The Cenotaph itself is visible proof of this confusion. Alfred Brookes, with an 'e', is listed on the First World War roll of honour and yet his half-sister Doreen Brooks is listed on the Second World War roll of honour without an 'e'. Does it matter? No, of course not, it's part of the magic appeal of the monument and the human story it is telling.

Alfred Brookes

Alfred Edwin Brookes (his definitive birth registration names) was born in Oakwood on 30th April 1890 and baptised in St. James Church on Wednesday, 17th September. His father was Alfred Brookes, sometimes called Alf, sometimes Fred. He was a collier. His mother was Minnie Brookes, née Baldwin. Alfred's service record has not survived but he was probably a Special Reservist and was on the Western Front at an early stage of the war. Alfred Brooks (name as per army record) was a private and his service number was 812. He was in the 3rd Glosters but he was drafted to the 1st Glosters, going abroad on 11th November during the horrors of the Battle of Ypres. When the Battalion were posted to Ypres in Belgium from France on 20th October 1914, they numbered 26 officers and over one thousand men. Four weeks later only two officers and less than two hundred men remained of the original unit.(1) Alfred himself was a casualty and his wounding was reported in the *Dean Forest Mercury* on 27th November. Interestingly, the newspaper report reflects the confusion over the surname:

> *'Mr Alfred Brooks, son of Mr. F. Brookes, Drybrook, Coleford has . . . been injured in the field . . . he is in hospital . . . '.*

The *Cheltenham Echo* of 16th February 1915 reported that he was wounded by shrapnel in the right leg. He was treated in hospital in Boulogne and subsequently Lincoln.

Between December 1914 and September 1915 the 1st Glosters were in trenches around Givenchy (France). Having returned to the Battalion, Alfred

Brookes met his death on Sunday, 9th May when the 1st Glosters took part in an attack on the Aubers Ridge at Fronelles. David Scott Daniell describes their '*advancing from their trenches into a murderous hail of bullets*'. They were recalled. Two hours later they went 'over the top' again and were again called back. In the two attacks the Battalion lost 12 officers and 275 men.(2) Alfred was one of them. He has no known grave and is commemorated on Le Touret Memorial near Neuve Chappelle. His family in Little Drybrook were notified of his death early in June.(3) As a youngster Alf had attended Ellwood Sabbath School and a Memorial Service was held for him at the Primitive Methodist Chapel in Ellwood in August.(4) He was killed nine days after his 25th birthday.

Alfred Brookes.
This is an enlargement from a photograph of Bream Junior Institute Rugby Football team, 1907, with assistance on identification by Don Ward and Jack Wildin.
Photo courtesy Jack Wildin

Leslie Main

Leslie Mayne, as he is on the Cenotaph, or Joseph Main, as he was known to the Army, appears in his birth certificate as Joseph Leslie Main. His name on the Cenotaph has the distinction of being the only one of the First World War names out of alphabetical sequence, coming as it does between 'Lucas' and 'Matty'. It is possible that this was a stonemason's error, because if the name were carved as Main then the alphabetical anomaly would disappear. Certainly the Cenotaph description indicates that in the village he was known not by his first name Joseph but by his second, Leslie. This was – and still is – not an uncommon occurrence in Bream.

Leslie Main was born at 104 Holmesdale Street, Lower Grange, Cardiff, on 1st August 1888. His birth certificate states that his mother was Alice Geraldine Main, his father is not named. His marriage certificate in 1912 does, however, name his father as Thomas Main (deceased), a former pilot. Leslie became a collier and as such was employed at Princess Royal. He was married on 26th December 1912 at Parkend Baptist Chapel to Hannah Preece of Rhondda, Mill Hill, Bream. The newly married couple lived on The Tufts and during his time in the village Leslie played full back for Bream Rugby Club.

He volunteered for Kitchener's New Army on 21st August 1914, became a corporal on 26th September and, after ten months' training, was sent to Gallipoli with the 7th Glosters, landing at Y Beach on 11th July 1915. He was killed at the battle of Chanuk Bair on 7th August 1915, when the Battalion was involved in a suicidal attack on a very strong Turkish position. The *Army and Navy Gazette* of 28th June 1919 reports that all 20 officers leading the Battalion were either killed or wounded and at the end of the day only 181 men were alive and fit for active duty, out of a Battalion 1,000 strong which had advanced that morning. On 15th August the remnants of the Battalion were temporarily amalgamated with the 9th Worcs. The official date of Leslie's death is 8th August, which is presumably the date it was confirmed by the stretcher parties. Ironically, his widow heard the news of his death from a Mrs. Ellis in Bream before she heard it officially from the War Office on 4th September. Mrs. Ellis' husband, a sergeant in the 7th Glosters, was a stretcher-bearer on the battlefield and reported home that Leslie had been *"shot through the heart". The Mercury* report of his death records that it *'cast a gloom over the village'* because he was of a *'cheerful and amiable disposition'* and had *'a large number of friends'*.[5]

Leslie held the rank of corporal in the 7th Battalion and his service number was 10399. He has no known grave. He was 27 years old six days before Chanuk Bair.

Leslie Main.
Photo courtesy
The Dean Forest Mercury

The village in 1915

The raising of a local battalion had been officially sanctioned before the end of the year and a public meeting was held in the village on Wednesday, 16th December 1914 to publicise its formation.(6) The force was to be a Pioneer Battalion not intended for front line active service except in an emergency. Their primary function was to carry out labouring jobs behind and in the line. Nevertheless, they were continually exposed to danger and 302 members of the Battalion were to die on active service on the Western Front.

The Mercury on 1st January 1915 contained a recruiting advertisement asking for 1,350 men for the Battalion. Interestingly, it contains a warning that men may have to take their place in the firing line if required. Sadly for many of them this was to happen, particularly in the spring of 1918. The idea of local battalions was enormously popular in Britain at the time and the Forest reflected this local pride. Recruits poured in to join the new Pioneer Battalion, which was to be called the 13th Battalion (Pioneer) The Gloucestershire Regiment. By 26th February 780 had joined, by 5th March 900 and by 19th March 1,130. Early in March the Battalion left for training in Malvern and in June for Yorkshire. On 17th December they were all granted six days leave on the assumption that the Battalion would soon be going abroad.(6)

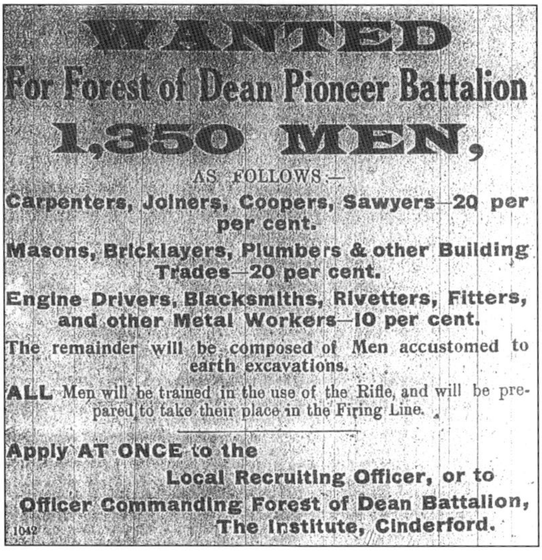

WANTED
For Forest of Dean Pioneer Battalion
1,350 MEN,
AS FOLLOWS:—
Carpenters, Joiners, Coopers, Sawyers—20 per per cent.
Masons, Bricklayers, Plumbers & other Building Trades—20 per cent.
Engine Drivers, Blacksmiths, Rivetters, Fitters, and other Metal Workers—10 per cent.
The remainder will be composed of Men accustomed to earth excavations.
ALL Men will be trained in the use of the Rifle, and will be prepared to take their place in the Firing Line.
Apply AT ONCE to the
Local Recruiting Officer, or to
Officer Commanding Forest of Dean Battalion,
The Institute, Cinderford.
1042

Forest Battalion recruiting advertisement, 1915.
Courtesy
The Dean Forest Mercury

It was not just the Forest of Dean Battalion that was recruiting locally. In the spring of 1915 a public notice appeared calling for a West of England Bantam Battalion. These were men disqualified on the grounds of height for service with regular units. The Bantams were recruited from men with a height of 5 foot to 5 foot 3 inches and a chest expansion of 34 inches. This West of England Bantam Battalion was to become the 14th Glosters.

On Monday, 23rd August 1915 the Bantams were recruiting in Bream. They paraded though the village behind a band and were given tea and a packet of Player's Navy cigarettes each at the Schools. After tea the Bantams marched to Sun Tump where a large crowd had gathered. The crowd were entertained by a display of drill on the cricket pitch. Then followed three speeches by two lieutenants and a sergeant calling on men to do their duty and volunteer. A report in *The Mercury* reflects the tactics employed and the mood of the occasion very well:

> *'The Force speakers all made very earnest and pointed appeals to the manhood of Bream, and the young ladies to do all in their power to persuade the young men to enlist for their country's honour. There were fully 500 people present.'* (7)

I do not know whether any local men actually did join the Bantams.

By mid-1915 it was clear that the supply of volunteers was inadequate for the requirements of the Army. In July 1915 the government passed the National Registration Act, which compelled everyone between 16 and 65 to register and by September the government had a breakdown of the population by sex, age and occupation. This showed that a large number of men not engaged in 'starred' occupations such as mining were still available. In October Lord Derby introduced the 'Derby Scheme', whereby men had one last chance either to enlist voluntarily or to attest that they would come forward if called-up at a later date. Finally, because half a million men were still evading any commitment to serve, full conscription was introduced in January 1916 for single men and widowers with no dependents aged 18-41. Conscription was extended to married men in May.

The National Registration Act led to a local couple from Saunders Green appearing at Coleford Police Court in August 1915. Thomas George Whitney and his wife, Mary Ann, were summoned for refusing to fill in the registration forms. After the enumerator had made repeated calls at their house with the forms to try and persuade the Whitneys to fill them in, Mrs. Whitney had (in the delicate language of the Court) *"made use of a filthy expression as to what witness was to do with them"*. Both were fined £1 and had to complete the forms. Clearly the Whitneys were worried about conscription, even though, as a collier, he was in a starred occupation. Their appearance before the Magistrates was not a good day for the Whitneys; on the same day Thomas was found guilty of keeping seven homing pigeons in his loft at Saunders Green without a permit. This was contrary to the Defence of the Realm Act. He was fined another £1!(8)

Given the horrors of the Western Front and Gallipoli, a reluctance to go to the Front was fully understandable. In October a near fatality occurred in the village when George Hughes, who lived in Lydney Lane, attempted suicide.(9) George was married with one small child. He played rugby for Bream and worked at the Princess Royal. In 1915 he volunteered for the Forest of Dean Battalion where his conduct was exemplary. In October he was on a short leave at home and, possibly depressed at leaving his family and what lay ahead, cut his throat. His wife, previously worried by his demeanour, had already sent for P.C. Head. The doctor was sent for and George's life was saved by the prompt arrival of Dr. Pugh. The Doctor himself was soon to leave the village for the Front.

1915 saw several presentations in the village as the people of Bream honoured their soldiers. On Saturday, 12th June a presentation evening was held at the Schools by Bream Amateurs A.F.C. for Jack Matty. The details of the presentation are recorded in Chapter 8. Jack subsequently returned to active service and died in Salonika in 1917. Also present that evening was another Bream footballer in khaki, Albert Thomas. Albert was to die in France in 1918.

In November Bream A.A.F.C. organised another presentation. This time a silver cigarette holder and case was presented to Gomer Cox, who was home on leave from serving in Gallipoli with the 7th Glosters. The following month the village honoured Charlie Miles, son of John Miles of Oakwood, who had been awarded the Distinguished Conduct Medal for gallantry earlier in the year.(10)

The Belgian refugees were then being looked after by the village and on 12th March a touching story appeared in *The Mercury*. '*Youthful Patriotism*' detailed how the children of the Infants School had been bringing in weekly contributions of halfpennies and pennies since the refugees arrived in Bream. Miss Young', the Headmistress, presented the money to the Belgian Refugee Committee Treasurer, Mr. J.H. Fewings.

At the end of July a Volunteer Training Corps company was started in Bream with the purpose of providing a Home Defence Force.(11) Till December 1916, they wore a lovat green uniform, when it was replaced by khaki. Their commandant was J.J. Dodgshon of Priors Lodge.

In some respects no doubt village life carried on as usual, i.e. with its capacity to throw up something unusual. *The Mercury* on 29th January records that a travelling family had been tenting in Bream for the last two weeks. They had with them a performing bear. Near the schools, the bear floored the man who was then rescued '*by the timely assistance of some residents*'. Clearly Bream's variation of a well-known Forest story is not 'Who killed the bears?' but 'Who rescued the bear's owner?'. What is it about forests and bears . . . ? Answers on a postcard please to Goldilocks . . . !

Hugh Royal does not appear on the Cenotaph roll of honour because he was living in Portsmouth at the time of his death in 1916 aboard *HMS Indefatigable*. Nevertheless, he was born in Whitecroft and living in Bream when he first joined the Navy, so merits inclusion here.

Photo courtesy *The Dean Forest Mercury*

1916

"*What Desolation In Homes War Makes . . .*"

(Joe Kent, 13th Glosters)

The year 1916 saw two massive battles on the Western Front, Verdun and the Somme. In February the Germans launched an attack on the French garrison of Verdun designed to *'bleed the French Army white"* (Falkenhayn, German Chief of Staff). Verdun was to claim over 700,000 French and German casualties.

Partly to relieve the pressure on the French, the British, under Haig, launched their own offensive on the Somme on 1st July 1916. This became the blackest day in the history of the British Army with nearly 60,000 casualties, most of them in the first hour of the attack.

The horror of the first day of the Somme has burned itself deep into the national psyche. No one from Bream died on that day but men from the village were certainly present and it is worth summarising one battalion's account of the battle to give some, albeit totally inadequate, idea of what happened. The Battalion is the 2nd South Wales Borderers, in which Ernie Lucas of Bream was a sergeant. The Battalion was attacking 'Y' Ravine, just south of where a mine had been tunnelled to explode under a German strongpoint – the Hawthorn Redoubt. The Battalion War Diary tells the story:

> *'07.20 Mine under HAWTHORN REDOUBT fired and Coys immediately commenced getting out of our trenches and through our wire . . . As the leading Companies reached the outer edge of our wire machine gun fire was opened on them which rapidly increased in intensity. Enemy also opened percussion shrapnel on the advancing lines. By about 07.30 the leading companies had lost nearly all officers and about 70 per cent. of the men . . .'*

The strength of the Battalion as it moved forward to the attack was 599. 399 of them became casualties, most of them within the first ten minutes. I do not know if Sergeant Lucas went 'over the top' that July morning because a cadre of 10 per cent of the Battalion had been held back, to protect against the whole Battalion being annihilated if the attack was a disaster. Ernie Lucas may have been one of the cadre. He was to die in 1918.

1916 also saw the only major sea engagement of the war, the battle of Jutland, which took place on 31st May. The British Grand Fleet was commanded by Jellicoe, who Churchill had previously described as *"the man who could have lost the war in an afternoon"*. Command of the sea was essential to Britain to supply her armies and protect her trade routes.

In the event the battle was inconclusive. The German Navy retired to port where it remained for the remainder of the war. On the other hand the Royal Navy lost more ships, including three battle cruisers, the loss of which was to provoke Admiral Beatty into the outburst *"There's something wrong with our bloody ships today"*. One of the battle cruisers *HMS Indefatigable* sank in similar circumstances to the *Hood* in 1941. Both ships were to suffer direct hits; both ships sank very quickly with nearly all hands (only two survived from the *Indefatigable* out of nearly 1,000); and both ships had a 'Bream connection'.

Hugh Mark Royal R.N. (280983) was aged 45 and a Chief Petty Officer (Stoker) on the *Indefatigable* when it sank at Jutland. He is not listed on the Bream Cenotaph but is commemorated on the Royal Naval Memorial at Plymouth (Panel 14). Hugh was born in Whitecroft and in the early 1900s lived at Yorkley. At the time of his death his residence was Portsmouth where his wife Frances and two sons, Will and Bert, lived.

Hugh joined the Navy in November 1894. At the time he lived in Bream, played centre for the Rugby Club and was captain of the team. Hugh had two brothers, Tom, who lived in Whitecroft, and George who lived at Windy Ridge, Lydney Road, Bream, both of whom also played rugby for Bream[1]. Hugh's obituary in *The Mercury*, 16th June 1916, is under the headline '*Bream Sailor drowned*'.

Having been decimated at Gallipoli in August 1915, the 7th Glosters were reorganised in Egypt and then sent to Mesopotamia. Here they took part in the relief of Kut, the occupation of Baghdad in March 1917 and the subsequent defeat of the Turkish Army. The campaign in Mesopotamia was to cost 29,000 allied lives, one of which, W.H. Pearce, is listed on Bream Cenotaph.

William Henry Pearce

William Henry Pearce was born in Parkend on 7th April 1897. His birth certificate registers his name as Peirce but the Vicar of Parkend records the spelling of his name at his Baptism in Parkend Church on 27th May as Pearce. This spelling was subsequently used by the Army and is the one used on the Cenotaph. His father was Henry Pearce, a coal miner. His mother was Emma Pearce (née Fields) who came from Aylburton.

At the age of 18, and presumably by then living and/or working in Bream, William joined the Army. He was a private (No. 26310) in the 7th Glosters serving in Mesopotamia. From 15th to 19th December the Battalion had been in the front line south of Kut. During that time their casualties were nine killed, nine died of wounds and 90 wounded. William died of his wounds on 18th December 1916, one week before Christmas.[2] He is buried at Amara War Cemetery, Iraq (grave reference X.A.2). He was 19 years old.

In the village in 1916

The battles of 1916 had a considerable impact on the village. The casualties for the Forest were truly awful and nearly every week *The Mercury* was printing obituaries of the fallen. The Town Hall at Lydney had been turned into a hospital. On 21st July, three weeks after the start of the Somme, a special fund was launched to cope with the wounded at Lydney. The question of setting up a hospital for the wounded soldiers at Bream was actively considered but discounted, according to *The Mercury*, on grounds of '*the water question and the sanitary conditions*'.[3] Instead, at a meeting at the Nationals Schools on Saturday, 25th November, the village decided to set up a committee to finance and administer The Lydney Craft School (now the C. of E. Primary

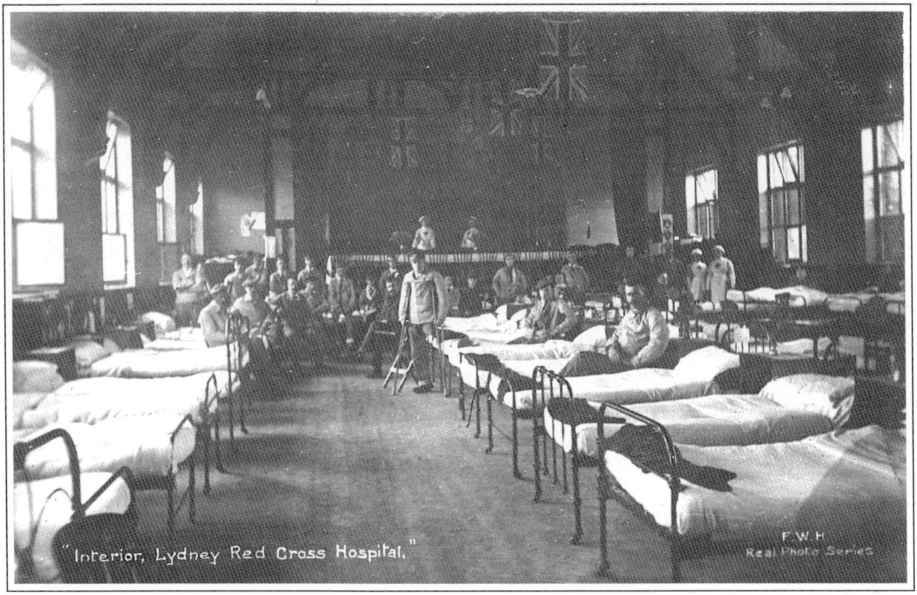

The Red Cross Hospital at Lydney Town Hall, looking towards the stage. There were four rows of beds for the wounded. Note this was prior to the building of the extension on the side of the Hall. *Photo courtesy The Archive Shop*

School) in Bream Road, which had recently become a hospital. The elected chair of the Committee was Rev. Cass, Mr. E.W. Morgan was Secretary and Milson Jones, Treasurer. Bream was divided into districts, each with its own collector who would call upon every household every week to raise money for the upkeep of the hospital.

The Forest of Dean Battalion, like many of the Kitchener Battalions, had gone on active service in 1916. They left Southampton on the evening of 3rd March and by the 11th they were doing typical Pioneer Battalion work on the Western Front '*constructing strong points, reclaiming old trenches, building dugouts and making breast works . . .*'.[4] The Battalion was involved in the Somme battle; for example, the War Diary entry for 3rd September 1916 records:

> '*A, B and D Coys went up to clear our trenches which were blocked by shell fire and corpses. All companies subject to five hour bombardment with gas shells and H.E. Little work possible . . .*'.

In November, the Battalion moved into the Ypres area in Belgium.

It is generally well known how difficult, as well as dangerous, life in the trenches was. An article in *The Mercury* on 21st January 1916 speaks of a plague of rats in the trenches. Terriers were effective against them and trainloads of dogs were despatched to the Front. A reward for every dead rat led to 8,000 'bodies' being handed in by one Army Corps (about 80,000 men) in one fortnight. As well as the rats, water in the trenches was a constant problem – not to mention the enemy!

There are a number of eye-witness accounts describing life in the trenches and, from time to time, new sources do come to light. No doubt most Bream soldiers sent letters home and we are very fortunate in the village that one soldier's letters have survived. That soldier was Joe Kent of the 13th Glosters (Forest of Dean Battalion). Joe was to die in 1918.

The three of Joe's letters to have survived are a mixture of a longing for home, a feeling of revulsion at the reality of war and, underlying all, his humanity and sense of humour.

The first letter, dated 6th September, is addressed to his sister, probably Julia. At the time the Battalion had been in action with the 39th Division as part of the battle of Guillemont which was, in turn, part of the battle of the Somme.

> '*It has been fearful here, hell is not a patch to it. Some of our fellows got knocked out. We lost our platoon commander, a nice fellow he was too. It's horrible to hear the shells screaming through the air, the whole earth seems to tremble. Never was there such a war as this before.*
>
> *I hope Daisy is better by now. I thought I would just drop you a line*

to let you know I'm alright, so is Ernie Evans [a friend].'

The second letter addressed to his sister Julia is dated 28th November 1916. The 39th Division had come out of the line on the 15th November and the battle of the Somme had finally finished on the 28th.

> *'. . . Well we shall soon have Xmas here. Have you made the puddings yet? I expect it will be a poor Xmas at a good many homes, many a mother's heart will be filled with sorrow this time, many a child will have to go short of a copper that he expected from his dad. What desolation in homes war makes, all through the Kaiser and his wicked clique . . .'*

The third surviving letter, reproduced in its entirety, is an undated (unattributed) transcript of a letter to his younger brother Fred.

> *Dear Fred,*
>
> *I am very well in health bar the cold. Its very cold here. I had just read in one of the London papers how the soldiers enjoy themselves in France behind the firing line, yes we get some damned enjoyment! We sleep in bean haulm* [stalks], *the barn is lousy. Our enjoyment is catching lice, it reminds me of going to Bristol Zoological Gardens where I used to visit the monkey-house and watch them killing nits. There is quite a variety of them here. I known of three distinct breeds, small red ones, medium sized and mackeral-backs and don't the buggers fight, and tear pieces out of our poor backs. Oh for a tin of Tommy Harrison's the king of nit pastes.*
>
> *. . . undress in a lime kiln when I go home, and get the old girl to give me a damn good white-washing. If you ever come out here don't forget to stock your valise with a good supply of nit powder. Perhaps you don't believe this, that there are two men in hospital with broken thumbs from cracking these impudent little devils but such is the case. Then there is the rats, thousands of them running over your face whilst you try to get a wink of sleep, that is another of Tommy joys. If I had been Pharaoh, King of Egypt, God could not have sent a better plague upon me than I've got here. I wish the writer of those articles was out here having a little of our enjoyment. Some nights he would be able to imitate Charlie Chaplin, for I can assure you we kick up some antics in the night. We are on the move again now, another enjoyment behind the firing line. I have got enough weight on my back to crush a mule. I am just like a duck marching, trotting up a new cut wheat field. If we don't get much money, by Gosh we have some fun. Well Fred we shall soon have Xmas here. I hope you will enjoy yourself, not nit hunting or rat scaring, but with some good beef and pudding. I daresay we*

shall have an extra tin of Ticklers jam and an extra biscuit and two hours extra drill nit killing. I myself have killed some thousands. Of course it would not do to publish them in the casualty list or else the country would be up in arms!

I wrote to Jack last week. I wonder how he likes soldiering. It's a fine game in your own country but its different here I can tell you, for a man never knows a minute but what he will take a voyage into the unknown regions. We go in a hurry when we start. Well I will now say "Good night" to you, wishing you the best of health and a jolly time Xmas, hoping to hear from you all again soon. I remain your affectionate brother,

Joe.

Joe's letters were kindly provided by Rex Morgan.

And other news from the village ....

The call up of the last of the Derby scheme men led to Mr. Eaton leaving the Boys School on 31st May. He returned the following day as being *'fit for garrison duty abroad'* and therefore he was sent back home – for the time being.(5)

A newspaper report on 30th June announced that Leonard Cross had now joined the Army. Mr. & Mrs. Cross of Bream now had all six sons serving.

The Belgian refugees had left the village for London and the Committee was wound up. The small balance in the funds was transferred to St. Briavels Refugee Committee.(6)

Throughout the year the village was contributing money for the war through Church, Chapel, street and house collections, and from donations from events such as the Flower Show. In July a War Savings scheme was started in the Princess Royal and Flour Mill collieries. Miners were encouraged to invest in government war certificates. Interestingly, Mr. J.H. Fewings, J.P. speaking for the Company said that they were prepared to put the last 6d to every certificate a worker bought. Mr. Fewings himself offered to see to the clerical work and Milson Jones was Treasurer. It was co-operation such as this that was to finance the building of the Cenotaph at the end of the war.(7)

The film *The Battle of the Somme* was shown in Coleford at the end of September and was, no doubt, seen by many in the village. Advertised as '*the most wonderful film ever shown*', it was based on footage that had been shot on the battlefield in July. It was however a somewhat sanitised view of events, designed to improve morale at home.(8)

1917

"Some Fitting Memorial To Those Local Lads Who Had Given Their Lives"

In 1917, changes occurred in the composition of the allies. In November, revolution took Russia out of the war. In April, the U.S.A. entered on the allied side though it was many months before 'Uncle Sam' could make an active contribution.

In terms of fatalities Bream had fared rather better than most communities in 1916. However, this wasn't to last. 1917 was a bad year for the village with eight deaths. Six of these occurred on the Western Front, the other two in one of the continuing sideshows – Salonika.

Four of the Western Front fatalities occurred at Ypres, or 'Wipers' as it was known to the Tommies. Wipers was a very dangerous part of the Front as it was a salient or bulge into the German lines. As such, British troops could be fired on from the front, side or even rear. To compound the danger, the Germans largely held the high ground around the salient and any breakthrough they achieved in the small Belgian city could continue to the Channel and possibly some of its ports.

There were three major battles at Ypres. The third battle, usually known as Passchendaele which was its final objective, was launched to displace the German Army from the high ground and produced conditions more horrific than anything in the war. August was the wettest in many years and the bombardment turned the battlefield into a sea of mud. When the Passchendaele ridge was finally captured, British losses were over 300,000. It is said that Lt. General Sir Lancelot Kiggell, who was Haig's Chief of Staff, in a rare visit to the front when it was all over burst into tears and said *"Good God! Did we really send men to fight in this?"* (1)

Two men from Bream, Willer Kear and Albert Watkins, died in the battle for Passchendaele. Another, Leslie Lodge, was seriously wounded and was to die of his injuries four months later in 1918. Two others, both in the Forest of Dean Battalion, died in the salient in 1917. One, Ernie Evans, before the battle; the other, Gordon Nelmes, just after it had finished.

Ernie Evans

Ernest Stanley Evans was born on 20th October 1879, at Cinderford. His father, Albert Evans, is described as a tinman on Ernie's birth certificate. Later in Bream he was remembered as a 'saw doctor'.[2]

Ernie was a stallman at the Princess Royal Collieries. He was single and lived with his parents in Parkend Road (near the old post office). He enlisted in the 13th Glosters in March 1915 and was one year at the front. He was a private (No. 19370) and died on Wednesday, 14th March 1917. Ernie was buried at Vlamertinghe Cemetery, Ypres (Grave VI.B.12). He was 37 years old.

News of Ernie's death reached the village by way of a letter from his chum, Joe Kent, to Mr. Burdess, Ernie's brother-in-law, requesting him to break the news to Mr. and Mrs. Evans. Joe also requested the letter be published in *The Mercury*. It was, on 23rd March 1917, and, in accordance with Joe's wishes, it is here reproduced in full:

'Dear Mr. Burdess,

It is my painful duty to inform you that your brother-in-law, Ernest Evans, was killed last night. Only yesterday morning, it being my birthday, he wished me many happy returns of the day. I shall feel greatly his loss. He was my chum, and used to work with me; a quiet, inoffensive chap, always smiling, and fond of telling a joke or listening to one. I thought I would let you know, knowing his father and mother, who are my parents' neighbours, and that by writing to you, you could break the sad news to them. It may be some consolation to you to know that he was killed instantaneously, therefore he suffered no pain. I know it will be a great shock to his parents, to you, and to his brothers and sisters to read these painful lines; it is hard for me to write this, but it is my duty. Ernest was a good soldier, and gave his life for his friends and country. It is a hard blow for you all, especially his aged father and mother. I therefore sympathise with you all. I shall greatly miss him. We often used to talk about Bream and the old folks there, and wishing for the time to come when we should get our leave, so that we should see again the old faces we love and to tell them some of the sights we have seen on this and other fronts. No doubt you will get a letter in a few days from his platoon officer informing you about his death, also one from the War Office giving you particulars about his death, but I thought I would just drop you a line or two to let you know it was the bursting of a shell at the mouth of a sap that caused his death. Two others were killed by the same shell. I shan't be likely to forget my birthday as long as I live. We had a terrible shelling, and many of us had some very narrow escapes. I have no doubt that his company officer will have a little cross erected at his grave. I may say this; our officers take great care of their men's graves everywhere where it is possible. They give their dead a decent burial. We have some very good officers who look well after the men's comfort. I am sorry to have to write such painful news to you, and I know that Ernie had a lot of good chums in our regiment. So I hope, Mr Burdess, you will convey to Mr. & Mrs. Evans my heartfelt sympathy, for I shall feel the loss of Ernie very much, and I know that it will be a great shock to his many friends in Bream. Well, Mr. Burdess, I will now close this painful letter with my fondest love and greatest sympathy.

I remain, yours very sincerely,
Joe Kent (Ernie's old chum).
Good-night and God bless you all.'

Willer Kear

William Thomas Kear died in the battle of Passchendaele on Tuesday, 10th July 1917. He was a gunner serving with 312 Siege Battery, Royal Garrison Artillery (No. 136229).

'Willer' was the second son of William and Mary Elizabeth Kear (formerly Preest) and was born at Bream Eaves on 14th September 1884. He was married on 2nd April 1908 to Mary Taylor, also of Bream Eaves, at Parkend Baptist Chapel. The Taylor family had recently moved to Bream from Yorkshire and Willer and Mary's old house is still called Methley House in reference to the village of Methley, near Castleford, where Mary originated from. There is a story current in the family that Willer, prior to his wedding, actually cycled all the way from Bream to Yorkshire! Willer and Mary Kear had three children, two daughters and one son, Hilda, Iris and William. At the outbreak of war Willer sold oil and china from

Willer Kear, his wife Mary and their eldest child Hilda, late 1908 or early 1909.
Photo courtesy of Margaret Awde

home and delivered oil to different parts of The Forest. He is buried at Adinkerke Churchyard extension, Belgium (Grave 815). When he was killed in action, Willer was 32 and left a widow and three young children. He never saw his son who was the youngest of the three.(3)

Willer's death was remembered in a poem written by his brother-in-law, William Alfred 'Fred' Taylor.

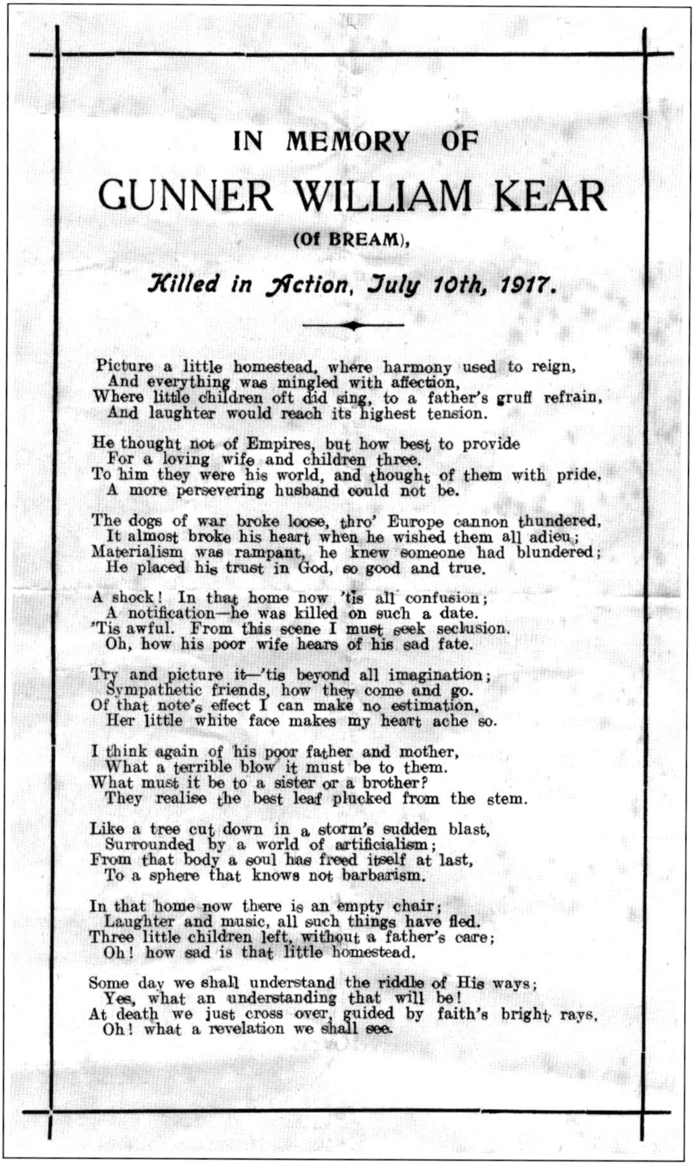

IN MEMORY OF

GUNNER WILLIAM KEAR

(Of BREAM),

Killed in Action, July 10th, 1917.

Picture a little homestead, where harmony used to reign,
And everything was mingled with affection,
Where little children oft did sing, to a father's gruff refrain,
And laughter would reach its highest tension.

He thought not of Empires, but how best to provide
For a loving wife and children three.
To him they were his world, and thought of them with pride,
A more persevering husband could not be.

The dogs of war broke loose, thro' Europe cannon thundered,
It almost broke his heart when he wished them all adieu;
Materialism was rampant, he knew someone had blundered;
He placed his trust in God, so good and true.

A shock! In that home now 'tis all confusion;
A notification—he was killed on such a date.
'Tis awful. From this scene I must seek seclusion.
Oh, how his poor wife hears of his sad fate.

Try and picture it—'tis beyond all imagination;
Sympathetic friends, how they come and go.
Of that note's effect I can make no estimation,
Her little white face makes my heart ache so.

I think again of his poor father and mother,
What a terrible blow it must be to them.
What must it be to a sister or a brother?
They realise the best leaf plucked from the stem.

Like a tree cut down in a storm's sudden blast,
Surrounded by a world of artificialism;
From that body a soul has freed itself at last,
To a sphere that knows not barbarism.

In that home now there is an empty chair;
Laughter and music, all such things have fled.
Three little children left, without a father's care;
Oh! how sad is that little homestead.

Some day we shall understand the riddle of His ways;
Yes, what an understanding that will be!
At death we just cross over, guided by faith's bright rays,
Oh! what a revelation we shall see.

Poem reproduced by courtesy of Roma Followes.

Albert Watkins

Albert also died in the battle of Passchendaele. He was a Lance Corporal (No. 241506) in the 2nd/5th Glosters and was killed in action on Wednesday, 22nd August 1917. His Captain, J.H.E. Rickerby, wrote to his widow:

> *'My Dear Mrs. Watkins,*
> *It is with very real sorrow that I have to tell you of your husband's gallant death. Your husband was killed instantaneously by a shell during the advance in which our brigade was taking part. I remember your husband well in England and feel the loss as that of a personal friend. He volunteered for every piece of work, however dangerous, and helped to form the backbone of the company. He was a splendid worker, and a most popular man in the company. There was no man in the whole battalion whom I admired more. He died near a famous town where a large number of British heroes lie buried . . .'.*[4]

Albert has no known grave. He is commemorated in the Tyne Cot Memorial (Panel 72-75) along with 35,000 other British soldiers who were missing in action in the salient after 15th August 1917.

Albert, born in 1889, was the second son of William and Polly Watkins, of Maypole, Bream. He was one of a family of eight children. Before the war he worked at Flour Mills Colliery. He married Kate Wicks at St. James on 9th November 1912. A son, Horace, was born in October 1913 and a daughter, Olive, in April 1915. Four months after Olive was born Albert volunteered for the Glosters and went to the Front in the spring of 1916. Albert was already something of a local hero in 1915 for assisting in the dangerous task of helping to recover a body from the Old China Pit in Oakwood Bottom.[5] He was presented with a mounted cigarette lighter and case for this. Edward Vaughan and Rowland Voyce

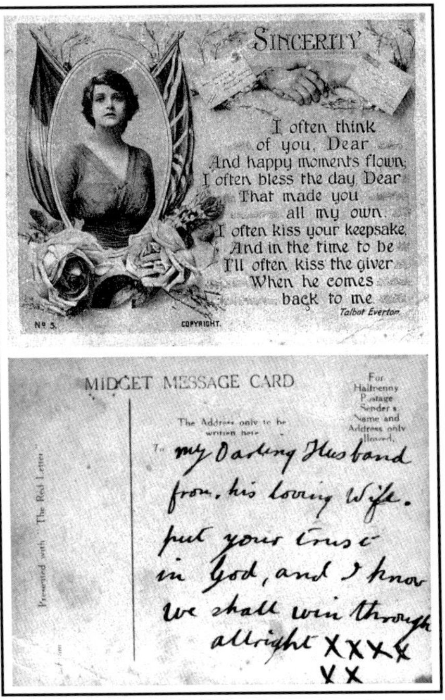

'Sincerity' – A 'Midget' postcard sent by his wife to Albert just before his death. It was returned to her with his effects. *Courtesy of Val Teague*

each received a gold medal for their work in that rescue.

Several of Albert's effects have survived. He sent a 'Dear Little Chicks' postcard to his children, whilst the 'Sincerity' message card was sent to him by his wife before he died and was returned to her with his effects. Albert had one home leave during his 16 months at the Front, which was possibly when the family photo was taken.[6] A poem in tribute to Albert, written by Joe Kent, has also survived.

At the time of his death Albert was 28 years old. His children were aged three and two.

Right: Albert Watkins, his wife Kate and their two children Horace and Olive.
Below: 'To my dear little chicks'. A postcard sent by Albert to his children.
Below right: Joe Kent's tribute to Albert.
All courtesy of Val Teague

"FORGET-ME-NOT."

To my Dear Little "Chicks."

A Soldier's Verse Letter to his children.

My dear little Chicks I'm thinking of YOU,
And dear Mother throughout the long day;
And wonder at times how you're getting on,
While your Daddy's so far away.

Now, be good children, and don't worry Mummy,
Then when I come home on my "leave;"
My "Kit Bag" I'll fill with Toys and Good Things,
Like Santa Claus on Christmas Eve.

Now, tell dear Mummy, I'd like to have,
A photo of you all—as you are;
Just to cheer me up while "doing my bit."
Though the distance between us is far.

I dreamt last night I was home again,
With you and Mother so dear;
But at early dawn the Buglers blew
"*Reveille*," in tones loud and clear.

FORGET-ME-NOT, my thoughts are of YOU,
While marching or from drill set free;
My loving thoughts for ever turn,
To Home—and Dear Ones—far from me.

From .... AT PARKHOUSE.

In Memory of Albert Watkins, of Bream.

He was a brave young hero,
He came at Duty's call;
In defence of home and Country
That brave young lad did fall.
Lance-Corporal Albert Watkins
Great sacrifice did make—
May we meet him in Heaven—
May he unlock the Gate!

Of that Celestial City,
Where God awaits us all,
When we have earthly Duty done,
And answered last Roll Call.
God bless his wife and children,
Whom he has left behind.
He was a noble Father,
To them he was so kind.

Upon the muddy fields of Flanders,
Killed by a German shell,
Lance-Corporal Albert Watkins
Fighting, bravely fell.
Bream is to-day the poorer
Through losing such a lad,
Whose parents, wife and children
To-day feel very sad.

No son could die a nobler death,
No more a lad could yield,
Than died this brave young hero
Upon the battlefield.
We all I'm sure shall mourn him,
And keep in memory still
The deeds that Albert Watkins
For England did fulfil.

J. KENT, 13th Glo'ster Regt.

Gordon Nelmes

Gordon Nelmes was born on 21st January 1897 at Bream Eaves. His parents were William Nelmes a collier and Harriet Amelia Nelmes (formerly Jones). He enlisted as a private (No. 19290) in the 13th Battalion early in 1915 and sailed with the Battalion for France in 1916. In October 1917, when home on leave, a concert was given in his honour at the National Schools and he was presented with a watch. He returned to Belgium and was killed on Wednesday, 5th December 1917.[7] His grave is located in Duhallow A.D.S. Cemetery, Ypres (Reference II.D.24). He was 20.

The two other fatalities on the Western Front in 1917 were Ted Kilby and Angus Dodgshon.

Ted Kilby

Edward Kilby, or Ted as he was usually known, was born on 20th July 1883 at Five Acres, Coleford. His parents were John and Matilda Kilby (née Aston) of Berry Hill. John was a collier.

Ted was one of a family of 12. He and five of his brothers (Tom, John, William, Henry and Sydney) were all in uniform at some stage or other. Tom Kilby was in the Grenadier Guards and served in the Sudan where he was wounded at the battle of Ondurman in 1898. He also served in the Boer War and had the distinction of being one of the guards at the lying-in-state of Queen Victoria in 1901. John Kilby was also in the Grenadier Guards and served in the Boer War where he was wounded at Belmont in 1899.[8] A third brother, William, was in the Monmouth Militia.

Three of the brothers served in the First World War. Henry was in the 13th Glosters (the Forest of Dean Battalion), Sydney was in the Royal Field Artillery and Ted served with the Monmouthshire Regiment and South Wales Borderers.

Ted Kilby was a keen sportsman and played rugby and cricket. In rugby he captained the Berry Hill Crescents, Cinderford and Bream, and there is a family tradition that he won a County Cap for Gloucestershire. This would appear to be confirmed by a newspaper reference circa 1910, lent to me by Roy Haviland, that two Bream players, Kilby and Saunders, may miss the Whitecroft game '*if selected for the county*'. He played in the forwards.

Ted married Gertrude Ada Bradley of Bream Eaves at the Primitive Methodist Chapel on 7th February 1910. At the time, Ted was a collier at Princess Royal where his father-in-law, William Bradley, was the Under Manager. In Bream they lived at Horseshoe Cottage, Parkend Road. There were to be four children of the marriage, Edward William, Annie, Ronald Ivor and Iris May, usually called Ted, Peggy, Ron and Iris respectively.

Ted was in the Territorials and decided to 'take the shilling' (a neighbour remembers this)[9] and go on active service in 1916. Possibly crucial in this decision was the death of his younger brother Sydney ('Strod') Kilby in the summer of 1916. Strod was badly wounded in an explosion in the July and was shipped back to Leicester Hospital, where he died. He was buried at Christchurch with full military honours on 9th August 1916. Ted attended the funeral and was shortly after drafted to France.

Ted was a corporal (No. 39932) in the 12th Battalion South Wales Borderers. On Saturday, 11th May 1917 news reached Ada Kilby that her husband was in hospital suffering from a gunshot wound to the chest. He died on Monday, 1st May 1917 and was buried in La Chapelette British and India Cemetery Peronne, Somme, France. His name is commemorated on the Roll of Honour of the South Wales Borderers in Brecon Cathedral, on Bream Cenotaph and the War Memorial in Christchurch, Berry Hill. He was also commemorated in a poem written by Frank Adams of Bream and printed in *The Mercury*, 25th May 1917.

Ted Kilby was 33 years old and left a wife and four young children. Ada Kilby subsequently remarried.

Edward Kilby with wife Gertrude Ada Kilby (nee Bradley) and first son Edward William, late 1910/early 1911.
Photo courtesy of Michael Kilby

Angus Dodgshon

In many respects Angus Dodgshon was different from all the other Bream men who died in World War One. For one thing he had more initials than anyone else (although the Cenotaph doesn't show this) and that in itself in the England of 1914 could be a rough indication of class and privilege.

Apart from his three initials there were other, more substantial, differences between Angus and the others on the Bream Cenotaph. Angus came from a world in which a son automatically had a private education and was expected to go to 'Oxbridge'. It was a world in which a London address and a desirable country residence were the norm, where parents were the pillars of local society and sons, if they went into the Army, would automatically hold a commission and not be a Tommy in the ranks. Angus probably didn't mix much with the boys of the village; most probably he was the envy of all since, rare privilege indeed, he owned a Norton motorbike.

Angus John Charles Dodgshon was born at Gilgerrau in Pembrokeshire on 21st April 1895. On Angus's birth certificate his father, John Julius Dodgshon J.P., lists his occupation as 'Gentleman'. Angus was educated at Hirst Court and Wellington College and was about to go to Trinity College Cambridge when war was declared in 1914. He then trained with the Inns of Court Officer Training Corps and was commissioned as a 2nd Lieutenant and, in December 1916, as a temporary captain. At this point in time he was 21.

The family had moved from Pembrokeshire several years before the war and lived at Priors Lodge, Bream, as well as having an address in London, SW7. Angus's father, John Julius Dodgshon J.P., seemed to occupy the role of 'squire' in the village. He is remembered for building the perimeter wall round the Priors Lodge estate to ensure greater privacy. Angus's mother was Ada Charlotte Dodgshon (née Dundas). John Julius was known locally as 'The Colonel' and when war was declared his leadership of the community was inspirational. In 1914 he inaugurated the Miniature Rifle Club at the Rising Sun, which would help the young men of the village learn to shoot straight. He became the Club's President and regularly shot in competitions. He was Chairman of the Belgian Refugee Fund, and chaired the village meeting in December 1915 which announced the formation of the Forest of Dean Battalion and appealed for recruits to the ranks. He was also made Commandant of the Volunteer Training Corps. His role in the village's war effort was, along with the vicar, the Rev. Cass, pivotal.

And yet John Julius only had one son; it is clear that the prospect of Angus being on active service worried him. An examination of Angus's service record highlights a situation that was stressful for all concerned but in its own way as fascinating as George and Mary Ann Whitney's refusal to complete registration forms in 1915, albeit this time played out on the much larger Whitehall, stage . . .

In May 1916 the 2nd/5th Glosters, in which 2nd Lieutenant Dodgshon was serving, were posted to France. Angus did not go with them because at an Army Medical Board he was passed as '*unfit for General Service*'. This verdict was based on a certificate (possibly instigated and paid for by his father), submitted by Dr. Maurice Craig, a Harley Street Nerve and Brain Specialist, who had been Consultant to Angus Dodgshon since 1911, when he was 16. At the time Angus was suffering from '*nervous exhaustion*' which had recurred subsequently '*owing to a concussion received in a Norton bicycle accident*'. The certificate concluded, despite Angus telling the doctor that he wanted to be passed for active service, '*I am strongly of the opinion that he ought to do Home Service only*'.

Consequently Angus was left at home when the 2nd/5th Glosters sailed for France. He was subsequently engaged on regimental duties in England and then he was attached to the 83 Prov. Battalion 12th King's Own Royal Lancers. By this stage he was a Captain.

'Disaster' struck in March 1917 when a further medical board at Coggeshall, Essex, passed him as '*fit for general service*'. Three days later his father, John Julius, wrote personally to Lord Derby, the Secretary of State for War and the Cabinet Minister who had given his name to the Derby Scheme in 1915, to ask for another medical board to reconsider the '*fit for general service*' label. I do not know if John Julius personally knew Lord Derby or not. The letter reiterates his son has '*nerve trouble*' and refers to '*rather an unsatisfactory family history*', and, in the absence of any further explanation in the letter, I can throw no further light on this comment. A full transcript of the letter, dated 30th March, is as follows:

To The R. Hon.
The Earl of Derby K.G.
Sec of State for War

41, Queen's Gate Gardens
S.W.7
March 30th 1917

Sir,

I beg to apply for another Med. Board examination for my son, Capt. Angus Dodgshon, who is at present in the 12th Battn King's Own R. Lancs Regt. He was passed as 'A1' by a Med. Board at Coggeshall Essex on the 27th March ult. but they then told him that they had no record of his medical history before then, as he had been moved fr. 2/5 Glos. to the 83 Prov. Battn. and then to the K.O.R. Lancs.

The previous Med. Board wh[ich] *he attended in the latter part of 1916 had before them the Enclosed Certificate fr. Dr. Maurice Craig, the Nerve and Brain Specialist, and they, in view of his medical history, passed him as unfit for General Service. He has since been engaged in various Regimental duties at home.*

The boy's general health is quite good apart from the Nerve trouble to wh[ich] *he has been subject. There is unfortunately rather an unsatisfactory family history.*

In the event of a further Med. Board being granted, I should be grateful if Dr Craig cd. be personally referred to, before any final decision be arrived at.

I have the honour to be Sir
Faithfully yrs,

John. J. Dodgshon

Two weeks later John Julius wrote a stronger letter to Lord Derby to request a further medical board '*as provided for in the regulations*'. Clearly nothing had happened to date and he was concerned his son would be sent abroad before a new board was convened. By this time the '*unsatisfactory family history*' referred to in the 30th March letter had become '*a very unfortunate family history*'.

To prepare evidence for another medical board Angus was again sent to see Maurice Craig who dutifully produced a second certificate concluding, '*I am very doubtful whether he will stand the strain of General Service*'. Unfortunately for John Julius, the consultant's recommendation in 1917 did not carry the same weight as it had in 1916. I do not know what Angus himself said to the Board, possibly he told them he wanted to go; he had said the same to Dr. Craig in October 1915. The Board's findings on 3rd July were succinct: '*He is fit for General Service.*' The Board ordered him to return to his unit and await War Office orders.

So Angus did finally get drafted as a 2nd Lieutenant to join his unit, the 2nd/5th Glosters, some 17 months after they had first sailed for France. The Battalion War Diary tells us that he joined them near Arras on Monday, 8th October 1917. He was killed on Saturday, 10th November when a shell burst ten yards from the trenches. He had been at the front for less than five weeks, barely time for Dr. Craig's doubt of whether he would '*stand the strain of General Service*' to be put to the test.

Angus is buried in the Sunken Road Cemetery at Fampoux in France (Grave reference I.D.27) as well as being commemorated on the Bream Cenotaph. At the end of the day one is left to wonder if there is any substantial difference between the protection Mary Ann Whitney gave her husband in 1915 and the protection John Julius Dodgshon gave his son in 1916 and 1917.

All of the differences between Angus and his fellow soldiers on the Cenotaph ultimately count for nothing. Fittingly, the Cenotaph has the final word on the matter with its democracy of death. Angus is not there as Lieutenant A.J.C. Dodgshon with his commission, his three initials and his Norton motorbike, he is there simply as A. Dodgshon, alphabetically between A. Brookes and E. Evans. In the end they are all equal. I like to think that John Julius Dodgshon approved of what was done on the Cenotaph. He had by

Angus Dodgshon with the other officers of the 5th Battalion Glosters at Chelmsford. Angus is sitting on the camera, left of the front row. The picture featured in *The Cheltenham Chronicle & Gloucestershire Graphic* of 27th February 1915.
Photo kindly provided by Graham Sacker

then moved from Priors Lodge to a house in Sussex but it would be a fitting epitaph of his enormous contribution to the village's war effort. He had, after all, lost his only son, who was aged 22 and had everything before him.

Angus is also listed on the Lydney War Memorial where he has been given an additional initial to those on his birth certificate. In Lydney he is recorded as H.A.J.C. Dodgshon.[11]

Away from the Western Front, a state of uneasy stalemate continued in Salonika, where two men from Bream died in 1917 – Jack Matty and Harry Wasley.

Jack Matty

John Matty, always known as 'Jack', was born in Tewkesbury on 2nd September 1886. His father was also called John and was a labourer by occupation. His mother was called Charlotte. As a young man Jack joined the Army and served in South Africa and India. After leaving the Army he came to Bream and worked at the Princess Royal. He married Florence Annie Collins in Tewkesbury in April 1914 and the marriage certificate describes his occupation as haulier. At the time Jack was playing soccer for Bream; the club then – as now – was known as Bream Amateurs A.F.C. He played centre half. His first match recorded in a local paper was against Lydney in the 1910-11 season.

Jack Matty, taken from a picture of Bream Amateurs F.C. 1913-4.
Photo courtesy of Jack Wildin

When war began Jack was immediately called to the colours and saw service in the 1st Battalion Glosters in the retreat from Mons, and battles of the Marne, the Aigne and first Ypres. He was wounded in May 1915, possibly in the same attack in which Alf Brookes lost his life. He was gassed and received four wounds, one of which was caused by a bullet through the cheek and tongue. He was invalided to Warrington Hospital and subsequently completed his convalescence at Bream, where he spent time with his wife Florence Annie and his new daughter Florence Annie, who had been born on 20th January and was baptised in Bream Church on 10th February.

On Saturday, 12th June 1915 a presentation function was organised at the Schools by the Football Club to honour Jack Matty before he returned to the front. The officials of the club organised a collection and the Doctor's wife, Mrs. Pugh, presented Jack with a silver cigarette case with the engraving '*Presented to Corporal J. Matty by the Bream A.A.F.C. June 12th 1915*'.

The vicar, the Rev. G.W.L. Cass, then spoke of his regret that Jack, who had suffered so much, should have to go back and it was a pity that someone could not go in his place. If he (the vicar) were 20 years younger he would himself offer to go in Jack's place.

Jack's wife, Florence Annie, probably had mixed emotions on that evening. There would be joy in being with her new husband (most of their marriage to date had been spent apart). There would also be joy at showing their five-month old baby and being together as a family after the suffering her husband had been through. And yet there would also be a dreadful foreboding. Jack would have to go back. Would he be as lucky next time?

Tragedy struck the family in May 1916 when the baby died of whooping cough and pneumonia. Less than 12 months later Jack himself was killed in action in Salonika. After the presentation evening in Bream he had returned to the Western Front with the 9th Glosters. At the end of 1915 the Battalion was posted to Salonika and here Jack died on Saturday, 17th March 1917 (three days after Ernie Evans). The Battalion War Diary for that date records very heavy shelling with '*between 4,000 and 5,000 shells*' falling on their section of the line. Jack held the rank of Lance Sergeant (Service Number 7831) and was buried in Karasouli Military Cemetery (Grave reference F.1358).

The Matty's lived in Box Tree Villa in Bream but as a family they had so little time together. At least they had their presentation night in the village, courtesy of Bream Amateurs A.F.C.

Jack was 29 when he died in Salonika.(12)

Harry Wasley

Henry ("Harry') Wasley was a son of Thomas and Maria Wasley (formerly Jenkins) of Mill Hill. He was born on 24th April 1896. In common with many of the men of the village, Harry worked at the Princess Royal. His father was an agricultural labourer (1881 Census).

Harry was one of the earliest in the village to volunteer and he served in the 11th Battalion of the Worcestershire Regiment, with the rank of Lance Corporal (No. 18303). He was killed in action on Wednesday, 25th April 1917, in Salonika. He has no known grave and is commemorated on the Dorian Memorial in Greece as well as on Bream Cenotaph.

At the request of his parents a Memorial Service was held in the Primitive

Methodist Chapel at the Eaves on Sunday, 27th June. Harry had attended the Sunday School and his parents were active members of the Chapel. The preacher was Mr. H. Cox.

Harry was just 21 years old. An In Memoriam notice appeared in *The Mercury* on 19th April the following year, 1918:

> *'He'll live forever in the hearts of those who loved him best.*
> *From his loving Mother and Father, Sisters and Brothers.'* (13)

Harry Wasley.
Photo courtesy The Mercury

In the village in 1917

The year 1917 saw the village very busy making presentations to servicemen. On Friday, 2nd March a concert was held at the Two Swans (owned by Mr. and Mrs. Wasley) and Private Leslie Lodge, who was home on leave, was presented with a purse of gold.(14)

At the Leslie Lodge presentation the idea of forming a Soldiers Fund Committee was discussed. This was formed in July and its officers were Mr. J.H. Fewings J.P., Chairman, Mr. W.F. Mullan the Schoolmaster, Secretary, and the Rev. G.W.L. Cass, Treasurer. During the rest of the year presentations were regularly made to servicemen on leave or those who had been invalided out. The presentations were usually a wristlet watch but not always so. Private H. Philips received a purse of silver, Private William Wintle a lamp and Private Alf Robbins a silver cruet. Dr. Pugh received a set of medical books. I wonder if any of these gifts have survived?(15)

The work of the Soldiers Fund Committee was important in the future development of the Cenotaph. As a Committee they decided to operate within the ecclesiastical Parish of Bream and not just the village. The Cenotaph similarly was to be based on the ecclesiastical parish and not just the village. The Committee also divided the village into districts each with collectors to raise money for the fund. This structure may well have contributed to the funds for the Cenotaph. Finally, at a special presentation concert on Thursday, 8th November a gift was given to every discharged serviceman. In moving the vote of thanks, Mr. Ralph Williams expressed the hope that the parish might set up '*some fitting memorial to the memory of all those local lads who had given up their lives*'. The idea was well received.[(16)]

A village rifle range was reported in *The Mercury* of 20th July. It was situated in Old Hang Quarries. A boy by the name of Scillet had a quite remarkable escape when he fell off the top into the deepest part of the quarry. He was doubly fortunate – to survive the fall and that the Rifle Range was not in use at the time!

At this stage of the war, London was being regularly bombed and – a foretaste of what was to happen in 1939 and 1940 – Bream hosted thirty children from the capital, who came for a holiday in July as respite from what was happening there. The instructions for their return journey were to meet at the Schools at 9am and then to walk to Whitecroft to catch the 9.30am train.[(17)]

Finally a nice example of 'pulling together' – the boys of the gardening class at School were digging the gardens of absent soldiers. The School Log Book records they were delighted to do their *'little bit'*.

1918

"To The Last Man . . ."

The withdrawal of Russia in 1917 released German troops for the Western Front. Determined to use this numerical advantage before American troops began arriving in large numbers, the German Commander, Ludendorff, launched a massive offensive against the British and French armies in the spring of 1918.

Ludendorff's offensive beginning on 21st March very nearly succeeded. The Germans pushed the French back nearer to Paris than they had been in 1914, whilst the British Army on the Somme was pushed back some 40 miles and the gains made in Belgium in 1917, at enormous cost, were lost. The situation was indeed desperate. On 11th April, Field Marshall Douglas Haig issued his famous backs to the wall order: "*There is no other course open to us but to fight it out. Every position must be held to the last man; there must be no retirement.*"

Eventually the German advance did run out of steam and on 8th August, labelled by Ludendorff as "*the black day*" of the German Army, the allies counter-attacked and achieved their greatest success of the whole war. This was the beginning of the end; on 26th September, the last line of German defence, the Hindenburg Line, was breached. On 9th November, the Kaiser abdicated and two days later, on the 11th hour of the 11th day of the 11th month, the armistice was agreed. The peace treaty ending the conflict was not however to be signed until June 1919.

The desperation felt in the British Army at the success of Ludendorff's offensive was reflected in local events. The 13th Battalion Glosters (Forest of Dean) was a Pioneer Battalion and was not trained as front line infantry. However, so desperate was the situation at the end of March that they were in the front line. Their losses were such that they ceased to be an independent battalion on 10th April; most of them formed a composite battalion with the 13th Sussex, others, the Headquarters Company, formed a cadre battalion to train American troops. On 11th May, this training cadre battalion became the 13th Glosters again but were officially disbanded as a fighting unit. They had more than done their bit.[(1)]

The desperation was also reflected in how individuals were treated. Percy Frowen, who had been discharged in 1916, was recalled to the colours on 10th April. Similarly, Mr. Eaton, who taught at the Boys' School and had not been passed for active service in 1916, was called-up and left school on 28th March.(2) From the allied perspective in 1918 the war was very nearly lost and then dramatically won.

For the village, 1918 was an even worse year than 1917. In just over ten months of fighting there were nine fatalities, eight on the Western Front and one in Salonika.

Leslie Lodge

Leslie is the only First World War serviceman who is actually buried in Bream. He was born Leslie Harold Lodge on 3rd July 1895 at Rose Cottage, Maypole, the second youngest son of Mr. & Mrs. Henry Lodge. Before the war Leslie worked at Downhams Bakery and the Princess Royal. He volunteered for the 13th Battalion (Forest of Dean) Glosters in January 1915. He subsequently served with the 10th Glosters and finally the 1st/6th Glosters. His service record has not survived but a report in *The Mercury* on 2nd November 1917 refers to him being previously wounded. We also know that he was home on leave when he received his purse of gold from the village in March 1917.

Leslie was fatally wounded on 9th October 1917 at Poelcapelle, during the third battle of Ypres, by a machine gunshot wound in the abdomen. The family account of this is that Leslie couldn't stand the smell of onions and when someone started eating onions in the trench that he was in, he got

Leslie Lodge (right) with Frank Downham, in the doorway of Downham's Bakery at Bream around 1913. *Photo courtesy Ruth Proctor Hirst*

out. He was then shot and, unfortunately, lay for a considerable time in a shell hole before he could be removed for medical treatment. Word came through to his parents that his condition was very serious and if they wanted to see him alive they would have to go to the hospital abroad where he was. This they did and stayed for a week. The patient seemed to rally and was brought back to be treated at Bradford War Hospital. He died there on 26th January 1918. Leslie was a private in the 1st/6th Glosters (Service Number 17986).

The onion story is a fascinating one and no doubt a true recollection of something that happened to Leslie. There are however two pieces of evidence that indicate he was possibly not escaping from onions at the time of his fatal injury. Firstly, there is his gravestone in Bream which states categorically that his fatal wounds were '*received in action*'. Secondly, there is the Battalion War Diary account of what happened on 9th October, the day he was seriously injured. Leslie's Battalion, the 1st/6th Glosters, went '*over the top*' that day in an attack on Poelcapelle. Zero hour was 5.20 am. The Battalion advanced against heavy machine gun fire and sustained 242 casualties. Many of the seriously wounded would have been in no man's land for hours – as Leslie was. In return the Battalion took 70 prisoners and captured 12 machine guns.

So if the onion story possibly does not relate to the events of 9th October, how can it be explained? We do know, as related previously, that Leslie had been wounded in 1916 or earlier in 1917. It is perhaps more likely that this is when the 'onion shooting' occurred. It is also possible that the smell of onions was in fact a smell *like* onions and that Leslie was the victim of a gas attack. It is certainly on record that there were times when British troops suffered from a British gas attack because the wind had changed direction.

This 'In Memoriam' card, complete with photograph of Leslie in uniform, was produced by his parents. *Courtesy Nancy Merry*

IN LOVING MEMORY

- OF -

OUR DEAR SON,

Pte. Leslie Harold Lodge

(13th Glo'ster Regiment),

Who was wounded at Ypres, October 9th, 1917, and died in War Hospital, Bradford, January 26th, 1918,

AGED 22 YEARS,

who gave his life for England, endeavouring to save us

Interred at St. James' Churchyard, Bream, Jan. 31st.

Leslie was brought home to Rose Cottage on 30th January 1918 and buried with full military honours in St. James' Churchyard on the 31st. Wenty Cooper told me that he remembered the road being covered with straw as a mark of respect to deaden the noise of passing carts.

This was Bream's only military funeral in the First World War. *The Mercury* estimated that a thousand people visited the churchyard, a mass outpouring of grief as the community mourned, not just for Leslie but all those others who had been buried overseas. The Bream and Lydney volunteers were present, and a firing party fired three volleys after the coffin had been lowered into the grave. The Lydney Silver Band were in attendance and a bugler from the Band sounded *'The Last Post'*.

In a sense every gravestone is unique. Leslie's is particularly so because it states that he was *'severely wounded at the Battle of Ypres in France'*. Ypres is, of course, not in France, it is in Belgium (Flanders). The mistake is perhaps symptomatic of a mood at that time which saw the Western Front as being simply 'over there' or 'across the Channel' and therefore assumed it to be France. This mistake on the gravestone somehow makes Leslie's death seem all the more sad.

At the time of his death Leslie was 22 years old. He was engaged to be married to Irene Farr of Berry Hill. On his last leave, possibly in the spring or summer of 1917, a considerate Leslie bought his fiancée's Christmas present – a brooch and a bracelet. An example to all young men who have been known to leave such things to Christmas Eve, assuming they remember at all! Possibly Leslie used his purse of gold from the village to purchase the gift?

Leslie's sister, Rose, described Leslie as having beautiful dark brown eyes – "*like velvet*". His photograph, kindly lent by her daughter, Nancy Merry, bears this out.[3]

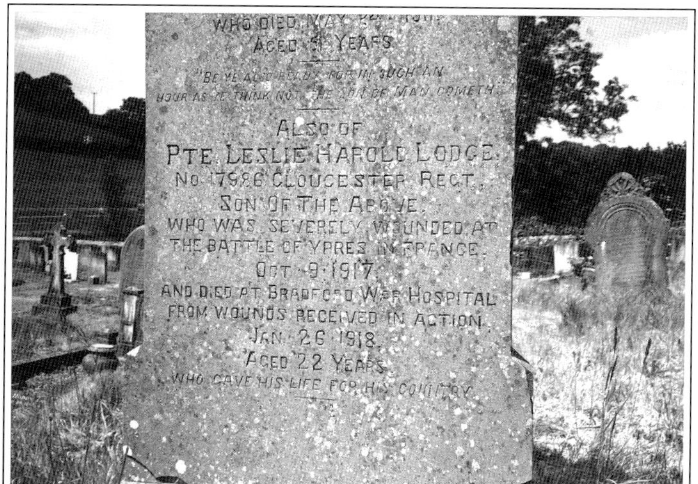

'Ypres in France'. Leslie Lodge's grave in Bream churchyard. *Photo courtesy Nancy Merry*

Albert Thomas
"At the going down of the sun and in the morning we will remember you."

Albert Edward Thomas is one of the two names on the 1914-18 Cenotaph who was decorated for gallantry. He was born in Bream in 1892 and was the eldest son of Rowland and Gertrude Thomas. He had three brothers, Frank, Les and Jess, and six sisters, Kate, Dolly, Victoria, Lil, Flo and Ede. Two addresses are given for Albert; the earlier one is Colliers Beech, Maypole and the later one is New Inn Cottage, High Street.

Albert volunteered in Bristol within a month of the outbreak of war. Fortunately, his service papers have survived, which – even though there is no photograph of him – give us a description: 5 feet 5 inches tall, 127 pounds weight, fresh complexion, grey eyes and light brown hair. His occupation was listed as miner and he was aged 22.

Having joined on 2nd September 1914, Albert underwent basic training and was then drafted to the 5th Battalion Middlesex Regiment (chosen probably because of family connections). His service numbers were 11007 and 26209. His entire service career was with this regiment but in different Battalions, the 5th, the 11th and finally the 8th.

Albert was posted to the Western Front on 27th June 1915. Two weeks earlier he had been in Bream and was present, as a fellow member of Bream Amateurs A.F.C. and also as a serving soldier, at Jack Matty's presentation evening. Four months after being posted he was wounded – '*shock by shell explosion*' – and was in the Casualty Clearing Hospital at Lilliers for several days.

In April 1916, serving with the 11th Battalion, Lance Corporal Thomas, as he now was, was wounded again, this time in an accident. He was hit in the calf by a splinter of a British Mills grenade. Some soldiers did self-inflict wounds to 'cop a blighty', to take them away from the front and back to England. There was an enquiry into Albert's injury since it involved him being shipped back to England. He admitted it had happened '*through stupidity*' when he had got on '*the parados of the trench to see the result*' (of the explosion). The central issue was whether it was the soldier's fault; the enquiry found that it was but it was an accident and not by intent, i.e. not deliberately self-inflicted. No action was taken against him.

The following year Albert was promoted to corporal and then sergeant in the 11th Battalion. He was awarded the Military Medal '*For Bravery in the Field*'. The M.M. was awarded to Warrant Officers, N.C.O.s and men. Albert's award appeared in the *London Gazette*, 18th June 1917: '*11007 Pte. (L./C.) A.E. Thomas Mdx R.*'

In February, Sergeant Thomas was transferred from the 11th to the 8th Middlesex. He died on Thursday, 28th March, having just been home on leave and presented with a wristwatch. Like many others he was killed in action resisting Ludendorff's offensive. He is buried in the Canadian Cemetery

No. 2, Neuville St. Vaast (Grave reference 16.8.2.) just north of Arras. He was aged 26.[4] The inscription on his gravestone chosen by his family reads: *'At the going down of the sun and in the morning we will remember you.'*

Gravestone of Albert Thomas, north of Arras, France.
Photo courtesy Eric Nicholls

Oliver Wilks

Oliver is the only soldier on the First World War Cenotaph who has a commemorative memorial in Bream apart from the Cenotaph. Parkend Road Chapel contains a plaque dedicated to his memory:

> *He was associated with this Church and Sunday School from early life having filled the offices of Society Steward and Sunday School Secretary.*

Oliver was born at Bream Eaves, now Montpellier Road, in 1888. His parents were Samuel Wilks, an insurance agent, and Elizabeth. He was the youngest of a family of seven. His elder brother Arthur was the father of Sam Wilks who died in the Second World War.[5]

Before and during the war, Oliver worked for Macfarlane's the grocer. The shop was at the top of Whitecroft Road but moved during the war to being opposite the chemist. The stables were at the rear of the shop. Oliver was frequently to be seen round and about the village with his horse and cart delivering groceries. The shop is still standing and the stables are currently undergoing building work.[6]

Oliver was conscripted on 1st September 1917 and was sent to France after a military training that was euphemistically described in *The Mercury* as '*not extensive*'. He was posted to the 2nd Battalion Royal Marine Light Infantry with the rank of private (No. 2417(S)).

Ludendorff called off his offensive against Amiens and the British Third Army on 5th April. The 2nd Battalion R.M.L.I. were in the Third Army and Oliver was killed on 6th April. He is buried in London Cemetery and Extension, Longueval, north-east of Albert and between the rivers Ancre and Somme. The grave reference is 7.E.23. Oliver was 33 years old.(7)

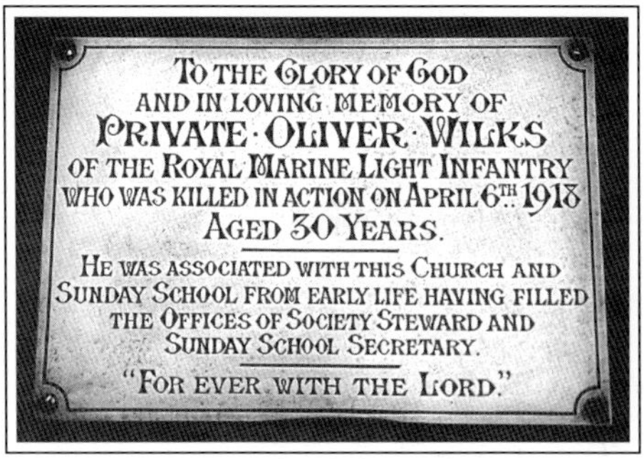

Left: Oliver Wilks
Photo courtesy David Wilks

Above: Memorial plaque to Oliver Wilks.
Photo courtesy Steve Cooper

George Winn

George Wynn's name as recorded on Bream Cenotaph is spelt Winn on his birth certificate and on his Army record. He is one of five Cenotaph names who are also listed on other local war memorials. He is listed on the role of honour in Coleford Church since, prior to living in Bream, he lived at Wynols Hill, Coalway. In Coleford Church his name is correctly spelt as Winn. Subsequently, the family moved to Collier's Beech, Bream, and then to the cottage opposite the church and facing the Maypole.

George Winn. *Photo courtesy his nephew George Batten*

George was born in Chancery Lane, Canton, Cardiff, on 19th March 1899. His father was John Winn, a mason's labourer. His mother was Mary Ann Winn. Mary Ann had been married before, as Mrs. Spiller, and was to marry again, as Mrs. Niblett, by the time she was living in Bream.

Before and during the first three years of the war, George was employed at Norchard Colliery. He joined the Army on 27th June 1917 and after basic training was posted to the Hampshire Regiment as a private (No. 132882). At the time of his death, on Friday, 26th April 1918, he was serving with the 4th Battalion, Machine Gun Corps.

News of what happened to George came to Mrs. Niblett in the form of two letters on the same day. One letter was from an Army chaplain at an advanced dressing station, informing Mrs. Niblett that her son was seriously wounded and was going to a casualty clearing station or hospital. The Chaplain wrote '*I promised to let you* [Mrs. Niblett] *know about him*', presumably on George's request to tell his mother. These may have been George's last words because the second letter written by a hospital sister states that George was unconscious:

> '*Dear Mrs Niblett,*
> *I sorry to tell you that your son, Private G. Winn, was admitted here at 9.15 this morning, dangerously wounded in the right arm, right eye, right thigh with fractured femur and tibia. He was very ill on admission and quite unconscious, and died 20 minutes afterwards (9.35). He will be buried in a little cemetery about two miles from here. With much sympathy.*'(8)

The '*little cemetery*' is Lapugnoy Military Cemetery, 6 kilometres west of Bethune, which now contains nearly 1,500 1914-18 graves. The grave reference is VIII.E.3.

George was 19 in the month before he died. His family believe that, had George survived the war, he intended to train for the priesthood.

Elton Worgan

William Elton Worgan was the second son of James Robert Worgan and Esther Worgan, of The Tufts. He was always known as Elton and is listed as E. Worgan on Bream War Memorial. Interestingly, *The Mercury* account of his death refers to him as Alton. Were they notified by a Forest accent?

By this stage of the war, with conscription in full operation, men no longer had the same choice of unit as they had in 1914 and 1915. Elton was posted to the 1st Battalion of the Lancashire Fusiliers as a private (No. 56797).

Elton was 19 on Wednesday, 12th June. He was killed the following day. His parents received a telegram on Saturday, 15th June that their son had been badly wounded in the back and leg. On Monday, 17th, another wire came stating that he had passed away.[9] Several weeks previously his elder brother Robert had been seriously wounded but was now recovering in Lydney Hospital.

Elton is buried in Bagneux British Cemetery, Gegaincourt, two kilometres south- west of the town of Doulleur, Somme. The grave reference is 111.8.31.

At 19 years and one day, he is the youngest of all on the Cenotaph.

Harry Howell

The H.L. Howells of Bream Cenotaph was born Harry Lambert Howell in the Forden district of Welshpool in 1893. He was the son of the late William and Sarah Howell. His father was a regular soldier who served 18 years with the South Wales Borderers. The family had Clearwell connections and settled there after William had finished with the Army. The family's itinerant Army lifestyle explains the Clearwell baptism of Harry's elder brother in 1886:

> *'William John, son of William and Sarah Howell, abode The Curragh'* [an Army camp in Ireland].

Harry Howell
Courtesy The Mercury

The family settled at Clearwell Post Office. During the war they were a famous local family because all of Mrs. Howell's six adult sons were in the Army. Four of the six volunteered for the 13th Glosters. A seventh son, Allan, too young to volunteer, was in the Coleford Boy Scouts!

The six serving sons were William, George, Harry and Bert Howell and their half brothers Ralph and Leonard Yarworth. The photograph of Harry Howell reproduced here is taken from a picture of four of the sons in uniform, which appeared in *The Mercury* on 16th July 1915.

At the start of the war, Harry was lodging with his aunt in Bream Woodside. He was one of six volunteers who put his name forward at the meeting in December 1914 to publicise the formation of the Forest of Dean Battalion. He served as a private in four different Battalions of Glosters; the 13th (No. 18086), the 1st, the 2nd/5th (No. 20400) and the 12th (No. 242379).

Harry was wounded in May 1917 and later returned to active service. He was killed in Flanders on Friday, 28th June 1918, aged 24, in an attack which cost the 12th Battalion over one hundred and fifty casualties. He has no known grave and is commemorated on the Ploegsteert ('Plugstreet' to the Tommies!) Memorial (Panel 5), 12 kilometres south of 'Wipers'.[10]

Percy Frowen
'A Soldier Called 'Frower' in a Place Named 'Sidney'.'

Percy's military career illustrates the depth of the crisis facing the British Army in the spring of 1918. He had been discharged in 1916 and placed on reserve but in the summer of 1918 he was recalled to the colours.

Percy was born in Monmouth Union Workhouse. His mother was Fanny Frowen, his father's name is not recorded on his birth certificate. When the war began, he was living with his wife, Alice Clara Kate Frowen, at 21 Sparrow Hill, Coleford. Because of this Coleford connection, Percy is also listed on the Coleford War Memorial. He and Alice had married at St. Peter's Church, Cheltenham, and had two children, Alfred William, born 3rd June 1915, and Gwendoline Mildred, born in Cheltenham on 2nd July 1917.

Despite the fact that he was a collier and would later have been in a starred occupation, Percy volunteered on 18th October 1915 and was, presumably at his request, posted to the Glosters. On 24th March 1916 he was transferred to 1 Gun Battery Royal West Kents and on 28th November 1916 he was discharged from active service and transferred to the Army Reserve.

The problem was a medical one. Percy, according to his service record, had '*rheumatical flat feet*', with the result that he was of limited use as a front line soldier and would be better employed working as a collier back home. Unless of course there was an emergency and, having trained him, the Army needed again to use him as a soldier.

The Army arranged for him to be employed at Princess Royal Colliery and he worked there from the end of 1916 until April 1918. The Army did keep in touch; on Percy's record there is a fascinating memo from the Princess Royal Colliery Co. to the Information Records Department at Hounslow. The

LYDNEY,
GLOS.,

[illegible] July 1917

Princess Royal Colliery Co., Ltd.

ALL COMMUNICATIONS TO BE ADDRESSED TO THE COMPANY.

Officer i/c
Infantry Records,
H O U N S L O W..

Dear Sir,

No 12786 Pte Frowen.P. Royal West Kent Regiment

Replying to your favour of the 23rd instant which though addressed to Sidney has now reached us. This man is still in our employ as a coal miner. Please also allow us to draw your attention to the name - you ask us about "Frower" but we take it this is a mistake and it should be "Frowen?"

Yours truly,

PRINCESS ROYAL COLLIERY CO. LD.

Above: The Princess Royal reply to the army's 'P. Frower who lived in Sidney' communication.
Courtesy The Public Record Office

Right: Percy Frowen.
Photo courtesy Gwen Warren

memo is in reply to an Army letter which had got Percy's name wrong, 'Frower' instead of Frowen, and thought he was working at the Princess Royal Mine 'Sidney' not Lydney. They also had his service number wrong – 12786 instead of 13786. Presumably the date of the Army letter, 23rd July 1917, was correct but one does wonder . . . . Blame the war!

The German spring offensive on the Western Front began on 21st March 1918. Percy was recalled to the colours on 10th April and told to report to the Royal West Kents' depot at Maidstone.

On a postcard, which has survived, Percy wrote to his sister, Annie Edey, just before leaving England:

> *August 7th*
> *Hello Sister,*
> *Journey down to Folkestone on train was fine. Just waiting now to board the boat.*
> *Kiss the children for me and don't worry about me. Love from your brother.*
> *Perce*

Percy joined the 12th Battalion R.W.K. in France on 12th August. He lasted less than two weeks at the front and was killed on 24th August 1918. He held the rank of private (Service No. 13786 – formerly 15939) and was aged 30. He is buried in Becourt Military Cemetery, two kilometres east of Albert in the Somme district.

Percy's wife Alice and two children, aged three and two, received a pension of 2s.5d (£1.27) a week. His widow also received his British War and Victory Medals, a plaque, a memorial scroll and his personal effects, which were his soldier's disc, a letter, a postcard and a photograph. The widow was now residing at 9 Townsend Street, Cheltenham. Other next of kin were his mother, now Fanny Mitchell, Bryn Cottage, Gilfach Road, Tonypandy, South Wales and his sister Mrs. Annie Edey, 5, Gloucester Road, Coleford.(11)

Joe Kent
"Mother Never Bred a Jibber"

Joe Kent, although he died in 1918, is still remembered in Bream as 'a character'. I was told he was an old soldier and spent time looking for Joe's service record prior to 1914. I didn't find one but his record may not have survived. In point of fact, he was certainly an old soldier in the very literal sense of the phrase; at the time of his death in 1918 he was 48 years old.

Joseph Kent, or 'Joe' as he was invariably known, was born in Upper Llanfrechfa, Cwmbran, on 14th March 1870. His father was Charles Kent.

His mother was Phoebe, who died in Bream on Boxing Day 1939 at the ripe old age of 92. She was Bream's oldest inhabitant at the time and is still fondly remembered in the village as 'Granny Kent'. Granny Kent was well known for her interesting reminiscences, including how she started work in the iron ore mines in Upper Cwmbran at the age of 9 and by 11 was working in a brickyard and could make eleven bricks a minute by hand![12]

Charles and Phoebe Kent had six children, of whom Joe was the eldest. There were three other brothers, Charles (Charlie), John (Jack) and Frederick (Fred). The two daughters were Julia and Elizabeth (Bessie). Three of the four brothers served in World War One, the exception being Charlie.

Like his father, Joe was a miner. In August 1909 he married Edith Robins, a widow from Pillowell. In 1914, it is possible to pinpoint some of his actions. After war was declared, Joe went in a pony and trap with Basil James and Basil's father Dapper to see the Territorials farewell parade in Coleford, on Wednesday 5th August 1914. Another story about him in 1914, related by Graham Field in *A look back at Norchard,* informs us that Joe, a miner at Norchard, made himself a recruiting office in the coal yard just off New Road, in a pile of pit props. He was busy trying to get the young men to sign up for the Army.

There are a number of references to J. Kent in *The Mercury* in relation to various activities in Bream. Unfortunately, it is unclear whether these references are to Joe Kent or to his brother Jack. Dare I mention that J. Kent scored one run for Bream in a notable game of cricket in August 1914? Lydney Athletic were all out for 26, in reply Bream were all out for 19. It was obviously a very close match – Bream only lost by seven runs! And how long did the match take – possibly 90 minutes at the most! One hopes arrangements were in hand for an early opening of the bar. Apart from cricket references there are also rugby reports of J. Kent. In both cases it is more likely that they relate to Jack and not Joe, since Jack is still remembered as a prominent sportsman who played cricket and rugby for Bream. As an athlete, Jack was reputed to be an 'even timer' capable of sprinting 100 yards in 10 seconds. During the war, Jack served in the Coldstream Guards. He was shot through the hip and thereafter always walked with a stick but this did not finish his sporting career; he continued to play cricket for Bream, the only concession to his injury being that henceforth he always batted with a runner![13] There are also references to a J. Kent on the Committee of the Miniature Rifle Club and the Rising Sun; this could be Joe.

Strictly speaking, Joe, being 44 in August 1914, was too old for military service. It is likely that he lied about his age when he volunteered for the 13th Glosters at Cinderford Recruiting Station on 20th February 1915. Could it be that he volunteered in Cinderford, where he was less well known, rather than in Coleford or Lydney where – on the basis of his recruiting activities – he was almost certainly well known by the local recruiting officer, Colour Sergeant Reeves?

A more likely explanation of Joe's enlistment has recently come to my attention from Ida Kear. Ida is the widow of Donald Kear, who was Joe's nephew. Apparently Joe liked a drink, preferably several, and could be something of a hell raiser. Donald's story is that Joe was well and truly drunk when he enlisted. After all, 20th February was a Saturday and recruitment did go on in the pubs. Joe spent the night in a police cell and when he woke up the following morning he was told what he had done and given the chance to change his mind. He refused to do so, saying "*Mother never bred a jibber*". And so Joe ended up in the Army after a heavy night out.

From Donald and Ida comes another story about Joe. Donald was born in 1913 and was only five when Joe died. Donald, as a little boy, received pictures drawn by Joe in the trenches and telling him that, even though there was a war on, he sometimes did hear the birds sing (many men back from the Front said they never heard birdsong when they were there!)

Margaret Addis, a one-time neighbour, tells another story about Joe. As a little girl she remembers him as being musical, playing in Bream Band and also in a trio consisting of a concertina, drum and fiddle. Margaret remembers the trio, possibly having had a drink or two, sometimes entertaining the children on their way home from school in the afternoon. The trio played the popular tunes of that time and the children sat on a wall, at the top of what is now Cox's Way, and joined in the singing. Margaret particularly remembers Joe teasing her by playing and singing,

> *'Maggie the cows are in the clover*
> *They've trampled it since morn*
> *Go and fetch them Maggie*
> *To the old oak barn.'*

Of course the children all joined in, much to Margaret's embarrassment and Joe's delight! Joe could be a tease with a lovely sense of fun.

Having enlisted, Joe spent most of 1915 training with the Battalion, before leaving for France in March 1916. Before he left however, Joe gave Basil James a walking stick to look after while he was away. The arrangement was that Joe would give Basil 6d ($2^1/_2$p) for looking after the stick while he was away. Basil subsequently passed the stick on to his sister.

Joe could have had at least one leave from the front on 28th March 1917. His sister Julia married Sam Cox, of Cox's Way, and one of the witnesses was J. Kent – but was this J. Kent Joe or Jack? No doubt the Vicar knew and assumed everyone else did.

Along with the rest of the 13th Battalion, Joe was caught up in the chaos of the Ludendorff offensive at the end of March 1918. Necessity had forced the Forest Battalion into the front line and by the end of April their casualty figures for the last two months were killed, wounded or missing 26 officers and 599 other ranks. Of the three hundred and 20 listed as missing, some had been taken prisoner by the rapid advance made by the German Army.[14] Lance Corporal J. Kent (No. 18512) 13th Battalion Glosters was officially a prisoner. *The Mercury* records the event in a touching article on 5th July 1918:

> *'Prisoner in Germany – Our many readers will be pleased to hear that our old friend – Corporal Joe Kent – is still in the land of the living. His wife received an official card stating that he was a prisoner in Germany and quite well.'*

The International Committee of the Red Cross have confirmed Joe's status as a prisoner. A list from the German authorities, sent to the Red Cross on 21st July 1918, confirms that Joe was captured on 25th April 1918 at Kemmel near Ypres. The Battalion War Diary reports that No.2 Composite Battalion, of which Joe was a member, were '*holding the line*' on the 25th. On the 26th they were completely surrounded from 7.30am to 8.00pm. They then fought their way back to the British line but only one officer and 17 soldiers reached safety. These survivors of No.2 Composite Battalion were then amalgamated with No. 4 Composite Battalion of 39 Division Composite Brigade. Such was the hotchpotch state of the British Army in the spring of 1918.

Joe was interned at Lager Friedrichsfeld. This stage of the war was not a good time to be a p.o.w. The allied naval blockade, in force since 1914, was having serious consequences for German food supplies and in such circumstances prisoners would not be a high priority. Joe died on Monday, 9th September and was buried in Glageon Communal Cemetery extension, Nord, France (Reference II.m.10.). The cemetery was used by the Germans for allied prisoners and is 11 kilometres south-east of Avesnes. Back home, Granny Kent never believed Joe was dead. Sometimes she woke up in the night claiming to have heard his voice calling from the woods.

We are fortunate that not only did Joe enjoy writing but also that some of it has survived. Reference has already been made to three of his letters about life in the trenches. He also published a tribute to Albert Watkins and one to Sergeant Jack Griffin, the Whitecroft rugby player. These are, by their very nature, sombre poems but his letters do contain evidence of Joe the character, the man who could even manage a wry smile in the most adverse situation.

Two other poems of Joe's have survived. One was called *'The Triple Alliance'*, probably about Tommy Kear, Tom Yorke and Sam Cox. The other poem, called *'Private Yorke's Band'*, is worth reproducing because it was written by Joe on 21st February 1916, which was 10 days before he went to France. It is likely that the poem was specially written for his embarkation leave, which would be his last time in the village before going overseas. The poem has a charm which is typical of Forest humour of a bygone age. Incidentally, Tom Yorke did have a band, there is a further reference to it in one of Joe's letters.

The strike referred to in the poem was probably the miners' strike of 1912.

PRIVATE YORKE'S BAND

No doubt you have heard of Tom Yorke
A Drummer of a Band
That used to play fine music
To hear them it was grand
Some years ago there was a strike
At a colliery in Bream
For getting higher wages
That was the miners dream

So they went to one Banks Jenkins
The owner of the Shraves
To form a deputation
Those men the colliery slaves
They formed a strong committee
To meet a man named Cooke
They wanted higher wages
But Cooke he would not look

At Tom Yorke or his merry men
So he brought forth his Band
To try and charm his Lordship
With music that was grand.
The drum had only one end in
That mattered not to Tom
So he hunted up John Hawkins
Who was a blower strong.

So off they went and me and John
Who had a double B
Their mascot was a badger
Twas beautiful to see.
They fell in by Jack Dappers
Who lives in Bream Woodside
Their tune was Rule Britannia
And off they went with pride

John Watkins had the Badger
'Twas caught on Bank's Estate
But soon it met its fate
They took him down to Whitecroft

Into the Miners Arms
Where people stood in thousands
The music had such charms

They looked at Johnny Hawkins
Who blowed his double B
Who urged his hearers to be firm
And win the victory
Tom Yorke he looked so happy
When marching down the lane
Be true, my lads, be true he cried
We shall the victory gain.

The landlord Ernest Baldwin
Suggested one good thing
And gave the band a quart or two
And hoped that they would win
'Twas there they sold the Badger
The price was half a quid
Ernie Baldwin was the purchaser
He gave the highest bid.

Now here's good health to Mr Yorke
The leader of the band
And also Mr Hawkins
That good old blower grand
I hope someday to meet them
Up in those mansions high
Where instruments are made of gold
And all are taught to fly.

Feb 21/16
Private Joe Kent
No. 18562 C Coy 13th Glosters [15]

Sadly no one wrote a poem about Joe or, if they did, it has not come to light. No matter, in reading what he wrote we find out something about him. I am sure he still enjoys playing in Tom Yorke's band in the sky!

Ernest Lucas

Of all the soldiers on the First World War roll of honour, Ernest Lucas was the most difficult to track down. He was not born in Bream and his service record contains no mention of a Gloucestershire address; his parents had left the village and Ernest himself was married with a home address in Hereford. Nevertheless the story of Ernest Lucas, when it did come out, was well worth the wait. He is the only regular soldier on the First World War memorial. He rose through the ranks to become a Company Sergeant-Major and subsequently a Lieutenant. He also won the French Croix de Guerre.

Ernest Lucas was born at 2 Power Street, Battersea on 15th January 1887. He was the son of Hubert Lucas, a railway labourer, and Sarah Elizabeth Lucas (formerly Redshaw). Whence any Bream connection? Hubert Lucas had grown up in Bream, living on Mill Hill with his parents and was an iron miner. Changing his employment took him away from the village to Battersea, London. It is possible that the family did move back to the village but Hubert, before the end of the First World War, had moved again to Eastleigh in Hampshire where he was employed at the London & South Western Railway's Works. To my knowledge, Ernest had two brothers, Francis and Albert, and one sister, Blanche.(16)

I do not know whether Ernest ever actually lived in Bream or whether he just visited family here. When he joined the Army at the age of 18 in 1905, his father's address was still Battersea and his own previous military experience was with the 11th Bn Royal West Surreys, which was a local force to Battersea. Nevertheless, he opted to join the South Wales Borderers when he enlisted, which was in Brecon. Both facts are an indication of his being in this area at the time.

When he enlisted, Ernest was described as being 5 feet 5 inches tall, weighed 138 pounds, had grey eyes, black hair and a '*fresh complexion*'. In answer to the question 'religion?' he replied 'Wesleyan'.

Ernest had signed up for nine years initially with the S.W.B., which would have taken him to 2nd September 1914; however he opted in 1913 to extend his service from nine years to twelve, and in 1916 he opted to complete twenty-one years. His first posting was in Great Britain with the 2nd Battalion, in 1907 he went to India for three years with the 1st Battalion and was then posted to South Africa with the 2nd Battalion. The rest of his military service was in the 2nd Battalion, until he was commissioned in 1917 and attached to the 7th Battalion S.W.B.

In 1912, Ernest was posted to China with the 2nd Battalion and on the outbreak of war he served as part of the Tientsin Expeditionary Force. In 1915 the Battalion was at home for two months, before serving in Gallipoli.

Throughout his career in the Army, Ernest had gained recognition and promotion. Initially a private, he was a sergeant by 1912. The advent of war meant advance was more rapid; in the Gallipoli campaign he was promoted Colour Sergeant on 22nd August and Company Sergeant-Major on the 23rd. On 10th March, the 2nd Battalion went to France as part of the build-up to the battle of the Somme, where they got so badly mauled the first day. It was in France that '*8866 Acting Company Sergeant Major Ernest Lucas, 2nd Battalion, South Wales Borderers*' was awarded the Croix de Guerre by the French Government. The date of the receipt of the award was 29th September 1916. The recipient would have been mentioned in despatches for an act of heroism. A copy of the award was sent to his father, Hubert Lucas, now living at 208 Chamberlayne Road, Eastleigh, Hants.

In January 1917, Ernest was commissioned as a Lieutenant in the 2nd S.W.B. The reason for the recommendation was '*His power of command and the confidence of the men. Has proved himself a capable leader in the field.*'

Ernest Lucas met his death fighting against the Bulgarians in Salonika on 18th September 1918. For three years there had been sporadic fighting in Salonika but the principle enemy seems to have been illness. More British troops died from illness than were killed in action. The Prime Minister, H.H. Asquith, described Salonika as '*a wild goose affair*' in which nothing happened.

Suddenly, in September 1918 things did happen and Bulgaria was hustled out of the war in the space of a few weeks. They surrendered on 30th September and were the first of Germany's allies to do so.

Ironically, Ernest Lucas died 12 days before the Salonika campaign finished. A letter from his C.O. states:

> *'It does seem a tragedy that after serving all through the Gallipoli campaign and then in France he should meet his death practically at the end of the war out here.'*

To date Ernest Lucas had had a 'lucky' war. He had been wounded once in Gallipoli in July 1915 but otherwise he was unscathed. His death, caused by a bullet to the head, was instantaneous. His C.O. wrote to his widow:

> *'He cannot have suffered a moment's pain. I saw him directly after he had been hit and he was lying on the ground with that charming smile on his lips which was so attractive.*
>
> *His valour was unquestionable and had he come through the fight he would doubtless have been decorated . . . The spot where he met his death was on a big hill called Grand Couronac about one mile from the town of Doiran. I enclose this small photo . . .'*

Ernest's widow went through the agony of Army red tape over her husband's death. He died on 18th September but the Army did not officially confirm his death until January 1919. Till then he was officially 'missing in action'.

The problem was that the 7th S.W.B. had been forced to retire, so there was no official confirmation of her husband's death and no burial. Eva had letters from Ernest's servant and two officers expressing sympathy at her loss but no information from the War Office other than that he was missing.

The servant's letter contains an interesting comment on the reality of war:

> *'. . . When the Batt. retired your husband was left there, if he had any personal property on him I think the Bulgars would steal it. I think he only had a wrist watch and a diary . . .'*

In frustration Eva wrote twice to the War Office and enclosed copies of the condolence letters she had received from Salonika. Both letters are on black-bordered paper. She knew he was dead. Eventually, on 28th January 1919 a statement was taken from his servant to verify Earnest's death. It then became official. In a situation such as this it is difficult to believe how slowly the bureaucratic wheels go round.

Eva Lucas was living at Dry Bridge Villa, St Martins, Hereford. Subsequently she became Eva Williams of The Cottage, Hoarwithy, Hereford. There were no children. Ernest Lucas was 31 when he died. He was, however, 32 by the time the War Office pronounced him dead. He has no known grave and is commemorated on the Dorian Memorial, Greece.

The Village in 1918

The German submarine campaign led to meat rationing, which had a very direct effect on the village – the Schools were closed. For the week beginning 29th April, the teachers were all off timetable to carry out clerical work in connection with the meat rationing cards. Obviously, even in 1918, teachers were good at managing inadequate resources! The Mixed School was to close again on 11th November – not because of the armistice but because of a flu epidemic! (17)

Bream and District Soldiers and Sailors Presentation Fund Committee continued to raise money and present gifts to soldiers on leave, or as they were discharged or demobilised. In all cases now the gift seemed to be a wristwatch.

Recruitment of soldiers continued to be a problem; perhaps Haig's phrase '*to the last man*' was becoming a reality at home as available manpower dried up. In March, some categories of miners lost their exemption and in April men up to 50 years of age were encouraged to enlist for certain categories of military service. Times were desperate.(18)

Conscription had been applied since 1916. Those who objected would have to attend an Appeal Tribunal and justify their objection on grounds of conscience. One such case affecting a 31-year-old married collier from Bream Eaves was heard in May. The man's defence was that '*Thou shalt not kill*' was a Divine command and that, as an individual, he believed in '*the brotherhood of man*'. Captain Wood, for the Army, then asked *"supposing you saw German soldiers marching up the road towards Bream should you feel it was your duty to take up arms to oppose their coming?"* The man replied *"I am against the taking up of arms . . ."*

In the prevailing climate of the time, the man's action, much scorned by public opinion, took considerable courage. His application for exemption was refused but he was given exemption from '*combatant service*'.[19]

A strange case relating to the Forest was heard in Bristol Assizes in February 1918. A man who had served in South Africa had rejoined the colours in 1914, only to suffer shell shock and be eventually discharged. As his counsel said, *"His body and mind were broken in the hell at Mons"*. Subsequently, he married a girl from Cinderford but it then transpired that he was already married and had been for twenty-two years. He had simply forgotten. He was found guilty of bigamy but a sympathetic judge sentenced him to two days imprisonment (which he had already served). He walked free from the court. The judge remarked:

> *"This is the first time in my 40 years experience that I have ever heard of defence in a case of bigamy that a man was married without knowing it"*. [20] (Part of the human cost of war!)

As the war drew to a close, Bream set up a 'Comrades of the Great War' organisation whose primary function was to look after ex-servicemen, for example, in respect of pension applications. The guest speaker at the inaugural meeting in August 1918 was Angus Buchanan V.C. from Coleford.[21]

The Armistice on 11th November 1918, although it was the cessation of hostilities, was not officially the end of the war. The Peace Treaty of Versailles, with Germany, was not signed until 28th June 1919 and extensive peace celebrations followed in Britain on Saturday, 19th July. Unfortunately, the weather was bad in Bream and everything was postponed except the huge bonfire on Sun Tump which was witnessed by a large crowd of people.[22] When the Schools closed on 31st July for the summer holidays – usually four weeks – they did not resume till 8th September, as the log records: *'an extra week being granted owing to the conclusion of Peace'* – with a capital 'P'![23]

Now that the war was officially over, local communities turned their thoughts to how they would remember the fallen. Newland's War Memorial was dedicated in March 1920, followed closely by Blaisdon in June and Drybrook in November. The last locally to be unveiled was Cinderford's, in November 1923. Lydney and Bream Memorials seemed to move in tandem. Before the end of 1920, work had started on the Lydney Cross. The land was donated by Lord Bledisloe and the monument was to cost some £800. The workforce of R. Thomas & Co. had offered to have a deduction from wages made for twelve months which would, it was believed, raise nearly £400 towards the cost. Subsequently, four villas were to be erected behind the cross, the contract being given to Hawkins, Harris and Weaver of Tutshill.(24)

In Bream meanwhile, a committee had been set up to organise a permanent memorial. On 27th August 1920 the following notice appeared in *The Mercury*:

BREAM PERMANENT MEMORIAL

'The committee are most anxious that no name should be omitted or wrong initial placed on the Cenotaph. Therefore would parents or relatives of any man from the Ecclesiastical Parish of Bream, who has made the supreme sacrifice in the Great War, be willing to give their assistance by sending that man's name and initials to

Geo Wilkins, Hon Sec.,
Woodside, Bream, Glos.'

Emphasis is again placed on the Parish and not the village. It is significant too that initials only were specified – only three names of the First World War Cenotaph have two initials and this must have been the families' preference. Similarly, those with two initials who were known by their second name – e.g. W.E. Worgan appears as E. Worgan – because that is how they were called by their friends and family.

Six weeks after this notice appeared, the Comrades of the Great War held their annual supper at the New Inn Hotel, hosted by Mr. & Mrs. Etheridge. At the meeting, the Vicar told the comrades that he would shortly hold a special service in memory of the fallen. It was then suggested Bream Silver Band should be approached to take part.(25)

What followed really did set the pattern for years to come. On Sunday, 7th November 1920, the Comrades assembled at the Sun Tump and headed by Bream Silver Band marched to the Church. The role of honour was read during the service, the Last Post and Reveille were played and the National Anthem was sung by a very large congregation.(26)

A similar pattern was followed in 1921. On Sunday, November 6th, the parade assembled at the Schools at 2.30pm. Headed by Bream Band and under the command of General Tyler of Priors Meend, they marched directly to the Church. The Cenotaph had been built by this stage and it is highly likely that there was an 'eyes right' at this point in the parade. Nevertheless, the Parish Church was the focal point of the Armistice celebrations and not the Cenotaph.(27)

The Cenotaph itself was unveiled on the afternoon of Sunday, 11th September 1921. General Tyler unveiled the memorial and the Vicar, Rev. C.H.T. Wright, performed the ceremony of dedication. A procession of ex-servicemen led by Bream Band marched from the Schools to Sun Tump. There the choirs from the Church and Chapels led the singing of '*Oh God our Help in Ages Past*' and '*O Valiant Hearts*'. A collection was taken to defray the cost of an iron fence to protect the memorial, or rather its flowers and steps, from 'ewe know who'!

The funding of the memorial was entirely local. Such balance as remained in the Soldiers and Sailors Presentation Fund provided the nucleus. Funding was then matched between the Princess Royal and the miners, £250 each. I would assume that the miners' contributions, as at Lydney, was by an agreed deduction from wages over a period of time. A similar co-operation, in embryonic form, had occurred two years earlier when the company had topped up miners' war savings. The co-operation between miners and their employers was all the more remarkable in that both years saw industrial action. In October 1920 the miners had been out on strike for a week. In 1921 there was a lockout for 14 weeks. It seems a matter such as the Cenotaph would transcend any level of industrial disharmony.[28]

In addition, there was a house-to-house collection throughout the Parish, putting a minimum cost on the monument of some £750 (equivalent now to a sum in excess of £22,000).

The architects for the project were Messrs. Kennard & Kennard of Gray's Inn Square, London, the same firm who designed the Lydney Memorial. The stone was purchased from the United Stone Firms at Parkend and the monument was erected by Mr. H.J. Walker. Harry Walker was a notable local builder. His building yard was where the Hedgehog car park currently is. Other work he did in the village was to extend the School in 1927 and build a lot of houses in Parkend Road. He himself lived at Stowe, in the last house on the right (going from Bream) before the quarry and the Travellers' Rest. Both he and his wife are buried in the tiny graveyard of the Congregational Church at St. Briavels.

THE SECOND WORLD WAR

The Second World War Cenotaph Roll Of Honour

AT THE GOING DOWN OF THE SUN AND IN THE MORNING
WE WILL REMEMBER THEM

M. BATH

N. BAKER
W. CROTE
L. COURT
R. CARPENTER
T. CHALLENGER
W. DREW
S. DREW
D. DAVIES
E. GRINDLE
D. HANCOCK
D. JENKYNS
V. JAMES
P. MEEK
R. MOORE
C. MORGAN
C. MARSHALL
J. NEILSON
J. OVERINGTON
T. TURLEY
D. WINTLE
W. WILDIN
S. WILKS
DOREEN BROOKS
DOROTHY HANCOCKS

1919 - 1939
"A Truce For Twenty Years"

Some historians take the view that there is a very strong link between the First and Second World Wars, and that 1919-39 was but an intermission in a conflict that began in 1914 and was not finished till 1945. Certainly, the Peace Treaty of Versailles in 1919 created at least as many issues as it resolved – as Field Marshall Foch said when he heard the terms, *"This is not peace, it is a truce for twenty years."*

The shortcomings of the Versailles settlement and German bitterness towards the terms were exacerbated by American isolationism in the 1920s and 1930s, which left Britain and France with the problem of making 'collective security' via the League of Nations a success. This was a burden they were incapable of carrying, all the more so in the 1930s when the world depression struck.

Furthermore, the policies of the Dictators, particularly Mussolini in Italy and, after 1933, Hitler in Germany, seemed to prove that aggression paid. Mussolini's conquest of Abyssinia and Hitler's piecemeal dismemberment of Versailles, were acquiesced in by the Western Democracies' policy of appeasement. When Hitler attacked Poland on 1st September 1939 he was confident that Britain and France would not interfere. He was surprised when they did.

By 1939, the one development that would have prevented Hitler's attack on Poland was an alliance between the Western Democracies and Communist Russia. This did not materialise. In fact the reverse applied, Hitler and Stalin agreed a ten-year non-aggression pact, and, in secret, the partition of the state of Poland. The war began eight days later.

On 3rd September Britain and her ally, France, declared war on Germany. As the war progressed Russia, the U.S.A and other countries joined the Allies. Germany was supported principally by Italy and Japan, which formed the Axis powers.

The British Prime Minister on the outbreak of war, Neville Chamberlain, has been much castigated over the policy of appeasement in the late 1930s. It has been argued that Chamberlain's appeasement was more a cause of the conflict than Hitler's aggression and had he been firmer, e.g. at Munich, then Hitler might have 'got the message'. In Chamberlain's defence, his attitude that war should be avoided at all costs was shared by much of British contemporary thinking. As Clement Atlee, the Labour leader, said in 1935, *"another war will mean the end of civilisation"*, whilst Stanley Baldwin, Chamberlain's predecessor as Prime Minister, stated also in 1935, *"the bomber will always get through"*. The Committee for Imperial Defence calculated in 1937 that there would be nearly two million casualties in the first two months of a war, leading the Home Office to work out the need for twenty million square feet of timber per month for coffins. They concluded the answer would have to be mass graves and burning with lime.[(1)] Such figures give a human perspective to Chamberlain's appeasement at Munich.

For the purposes of this book, the war falls conveniently into four connected but distinct phases: The Phoney War – September 1939-40; Britain Alone – 1940-1; The Turn of the Tide – 1942-3; and The Collapse of the Axis –1944-5.

The Phoney War
September 1939 - May 1940

It was assumed that the declaration of war in Germany would be followed by a wave of mass destruction. In fact nothing of the sort occurred, Britain was busy preparing for total war but there was to be no action in Western Europe for eight months. The Germans sat behind the Siegfried line, the French the Maginot Line and the R.A.F. dropped propaganda leaflets rather than bombs on Germany. This period of suspended hostilities was called Sitzkrieg (the 'Sitting Down War') in Germany in ironic contrast to the Blitzkrieg (the 'Lightning War') that had destroyed Poland in a month. In Britain, people spoke of the 'Phoney War'.

In September 1939, however people assumed the worst. Everyone was encouraged to have an identity label and carry a gas mask at all times. A blackout was immediately enforced and a mass evacuation programme began to move some two million children from the vulnerable city areas to the countryside. All cinemas, theatres, dance halls and places of public entertainment were closed until further notice. Every community had its special constables and its air raid wardens. Men between 18 and 41 were conscripted and those up to 50 were encouraged to volunteer for the Home Defence Force.

In terms of population, Bream had grown little since 1914 with a population of 2,649 (1931 Census).[(1)] However the late summer of 1939 saw dramatic changes in the village, notably the arrival of refugees and the blackout.

The Mixed School Log Book entry for 5th September records:

> *'School was due to commence today but in view of the National Emergency in common with all other schools it remained closed. The Head Teacher was on duty taking the names of boys who had been evacuated to Bream, some under the Government Evacuation Scheme, some under private arrangement and some continuing their stay in Bream (as advised by the BBC).'*

On 6th September three evacuees joined the Infants' School and the following Monday were placed on the Evacuation Register. Fortunately this has survived and we know their names, one five year old, Walter Minor, and two six year olds, Yvonne Enright and James Duff.[2] As the Phoney War continued a lot of the original evacuees returned home. By February 1940 only Yvonne Enright was left on the Infants' Evacuation Register. A report in *The Mercury* on 6th December states that, of the original 308 evacuees to West Dean (93 mothers and 215 children), only 114 remained (20 mothers and 94 children – and most of these were unaccompanied children). The numbers were to drastically change again in six months' time.

In respect of the blackout the Forest had more than its fair share of hazards. Then as now Forest Roads contained four-footed pedestrians. In 1939 the blackout speed limit was 20 m.p.h. but there were no street lights and all cars had masked headlamps which gave out very little light. In the period 1st September to 31st December Forest Police reported double the number of road deaths as compared with the same period in 1938 (six as opposed to three).[3] The usual Forest black spots were even blacker and on 22nd September *The Mercury* reported a miraculous escape for a lorry driver on Blakeney Hill, driving down to '*Fatality Square*'.

Predictably the blackout was not popular, particularly as the Phoney War continued and there seemed less need for it. A number of cases of breach of regulation came before the magistrates and Bream seemed to have more than its fair share of offenders. One, Mr. Davey of Parkend Road, was so incensed with how he had been treated that he resigned from the British Legion in protest![4]

Food rationing began in Bream, as in every other community, in January 1940. Clothing followed in 1941.

Everyone was encouraged to be security conscious by a flurry of propaganda posters. Among the more famous were 'Careless Talk Costs Lives' and 'Tittle tattle cost the Battle'. Composing them became a pastime with the R.A.F. and one of my favourites is 'Why not be as dumb as you look?'[5]

Britain Alone
1940 - 1941

The war 'proper' began in Western Europe on 9th April 1940, when Hitler invaded Denmark and Norway. The prime purpose of the invasion was to safeguard German iron ore, which was shipped from Sweden via the Norwegian port of Narvik. At the time Britain was herself considering an attack on neutral Norway to disrupt these iron ore supplies. In the event, British Troops were sent to Norway in the role of defenders. The failure of the British intervention in Norway led directly to Chamberlain's resignation as Prime Minister on 10th May, when he was brought down by his own party, the Conservatives. One of his own backbenchers, Leo Amery, sounded the death knell for appeasement when he attacked Chamberlain and dramatically quoted Cromwell's dissolution of the Rump Parliament in 1653 *"In the name of God, go!"* Churchill then became Prime Minister of a coalition government.

On the same day, 10th May, Hitler unleashed Blitzkrieg on Holland and Belgium then, after skirting around the Maginot Line, invaded France on 12th May. Within six days the German advance had reached the English Channel. In the north the British Expeditionary Force (B.E.F.) and the French Army were compressed into a shrinking perimeter around Dunkirk. A heroic evacuation rescued over 300,000 troops and brought them back to England. Subsequently France capitulated on 22nd June. Britain and the Empire now stood alone.

The fall of France was the precursor for the Battle of Britain, which began over England on 13th August, a sad day for the village of Bream. Initially the Luftwaffe concentrated on R.A.F. fighter command airfields to gain control of the skies as a prelude to invasion. In September the German command blundered. The Luftwaffe switched their attacks from the airfields and on 7th September attacked London. This gave fighter command the chance to recover. On 17th September Hitler called off 'Operation Sea Lion' indefinitely; there would be no invasion that year. Nevertheless the winter of 1940-41 was a hazardous time to live in British cities. From 7th September, London was bombed for the next 76 nights, except 2nd November. Nearly all of Britain's major cities were subsequently Blitzed. By the summer of 1941 when the Blitz ended, two-and-a-half million people were homeless and 43,000 citizens had died.

In 1941 the entire character of the war changed. On 22nd June Hitler ordered 'Operation Barbarossa', the attack on Russia. On 7th December aircraft of the Japanese Navy attacked the American Pacific Base at Pearl Harbour. A few days later Hitler declared war on the United States, so that when the year ended, Britain had been joined by two powerful allies, Russia and America.

1940 was a bad year for the village. Between May and December there were six fatalities. 1941 was little better with four.

The first to die from the parish was Morris Bath, in the ill-fated Norwegian Campaign in 1940.

Morris Bath
'A Good And Gallant Soldier'

William Benjamin Morris Bath was born at Allastone Mesne on Sunday, 2nd April 1905. He was the only son of Ben Bath, Master Baker in Whitecroft and Bream, and Sarah Anne Bath (née Davies). After leaving Pillowell School, he worked initially in the family business with his father, mother and two sisters. Among his hobbies at the time was boxing, and the hook on which his punch-bag hung could, until recently, still be seen in the ceiling of the Barbers Shop at Bristol House. He was also keen on motor bikes and the included photo shows him proudly astride his 'Indian'. In 1928 he married Alice Woodhouse, of Redbrook, at Newland Church.

Morris Bath on his 'Indian' motorbike.
Photo courtesy Clive Bath

At the outbreak of war he was in the Territorial Army and working delivering coal in Hereford. He was called-up and was assigned to the 1st Battalion, D Company, the Herefordshire Regiment. After a short training period the regiment was sent to Norway. The major British offensive was directed against the port of Narvik but British troops were also landed to the south at Andalsnes, Trondheim and, some 250 kilometres south of Narvik, at Bodo. Morris's regiment initially landed at Trondheim as part of 'Bodoforce' but were subsequently based at Bodo. During the withdrawal from the village of Finneid on 26th May 1940, he was severely wounded and taken by boat to a Hospital at Fauske where he died the following morning. This is confirmed by a letter from his C.O. (addressed incidentally to Mrs. M. Bath, Bristol House, *Brein*, nr. *Sydney* – another variation on our address theme!). A letter from the medical officer looking after Morris confirms the time of his death as 5.15am on 27th May and touchingly says of him:

> *'I should like to tell you how brave and unselfish he was – he was most grateful for all that we could do for him – we all felt that a good and gallant soldier had passed away.'*

Orders to evacuate 'Bodoforce' were issued on Sunday, 26th May and by 29th May the withdrawal had been completed. The attendant confusion probably explains the error on the original headstone, which (and listed as such in the C.W.G.C. records) states Morris died on 26th May. In fact he died on the 27th. He was buried by the British, just prior to the evacuation, in the small cemetery at Fauske but it was left to a grateful Norwegian public to erect a headstone. They chose a granite stone 1.5m high with the following inscription:

PETE BATH
BRITISK SOLDAT
FALT UNDER KAMPENE
I SALIDAL 26 MAI 1940
TAKK FOR DIN INNSATS

The British identification of the grave would have listed Morris as Pte, i.e. Private, Bath, which the locals obviously took to mean that his name was 'Pete'.

A full translation of the rest of the inscription would read:

BRITISH SOLDIER
KILLED IN BATTLE
IN SALTDAL 26 MAY 1940
THANK YOU FOR YOUR BRAVENESS IN DUTY [(1)]

Morris is the only British soldier buried in Fauske Churchyard. He was a private (No. 4105644) and was 35 at the time of his death. He left behind a widow and two young children, aged seven and 18 months. The original lettering on the gravestone has since been altered to read 'W.B.M. Bath'.

When Morris's widow Alice May Bath died aged 90 in 1995, her last wish was that a replica headstone, the exact shape and size as in Fauske Churchyard, be erected on her grave in Bream Churchyard. This was done. The headstone in Bream also commemorates Morris.(2)

Top left: The original headstone in Fauske churchyard, erected by the Norwegians to honour 'Pete' Bath.

Top right: The correct headstone in Fauske churchyard.
Both photos courtesy Clive Bath

Right: The replicated Fauske headstone in St. James's churchyard.
Photo courtesy Ruth Proctor Hirst

Bill Crote
'God Moves in a Mysterious Way'

Bill Crote was in the R.A.F. Volunteer Reserve and was called-up following the evacuation from Dunkirk in the late spring of 1940. After the fall of France in June, the Luftwaffe mounted a series of attacks on British convoys, not merely to sink ships but also to lure R.A.F. Fighter Command out over the channel, where a pilot shot down might well be a pilot lost and where superior numbers of short range German Messershmidts could operate easily from airfields in occupied Northern France. British historians date the beginning of the Battle of Britain as 10th July; German historians however use 13th August as the start date. This was 'Adlertag' or 'Eagle Day', when the Luftwaffe attacked the airfields of R.A.F. Fighter Command in the South East prior to the full invasion of Britain, scheduled for 15th September.

Aircraftsman 2nd Class Bill Crote (No. 1010974) died at R.A.F. Detling on the first day proper of the Battle of Britain, 13th August.

Bill Crote.
Photo courtesy Joan Lee

William James Crote was the next but youngest in a family of seven and was born in 1913. His parents were John and Maria Crote of Mill Hill. His mother, who was always known as 'Tiny' in the village, died in 1923 when Bill was nine years old. The cottage where he was born has since been demolished and a modern bungalow, Ty Bryn, now stands in its place.

Bill was educated in Bream and was known as 'Crotey' at school.[3] In the mid-1930s he married and settled in Birmingham, although his parents still lived in Bream. Bill's first posting in the R.A.F. was to R.A.F. Detling near Maidstone in Kent.

The first raid on Detling took place just after 17.00 hours on 13th August, when the Stuka Ju 87s of Luftflotte 2 attacked. It was teatime and the mess halls were filling with personnel, 67 of whom were killed. Twenty-two aircraft were destroyed on the ground and the Operations Block was devastated. Ironically, R.A.F. Detling was not in fact a Fighter Command base.[4]

Bill Crote was killed in the raid. He had been in the R.A.F. for less than six weeks. His father met his coffin at Lydney Station to bring him home for burial. There was a note attached to it stating that under no circumstances was the coffin to be opened.[5]

Bill was buried with full military honours in Bream Churchyard on 20th

August 1940 (Row E, Grave 29). His headstone has the epitaph:

"God moves in a mysterious way"
Loving memories
From his wife

His wife, Winifred Iris Eileen Crote, spent her remaining years in Coalway. Bill was 27 when he died.[(6)]

Len Court
A 'Fitter' in Peace and in War

Len Court also died in the Battle of Britain, killed in a bombing raid on Bristol. Of all the names on Bream Cenotaph, his was one of the most difficult to trace. He was not born and bred in Bream, had left the village by the time of his death and has no relatives in the locality. He also, for some reason, is currently unlisted on the C.W.G.C. role of honour of civilian casualties.

Leonard George Court was born in Bristol in 1906, the son of Albert James and Laura Izett Court of 7 Meridian Vale, Clifton.

Len came to Bream to work. He was employed by F.W. Mullan as a dental mechanic. The dental surgery was situated in the building which is now the Hedgehog public house.[(7)] It is interesting to speculate how many sets of false teeth Len made for people in Bream!

As a young man, Len lodged at Wellmeadow, down at the Maypole. He also played football for Bream.[(8)] Wenty Cooper, was responsible for introducing Len to his future wife. A group of four young men had gone by car to a Saturday night dance at Broadwell. The dance was poorly attended so they got back in the car and went to another dance, this time at Brockhollands. At that dance Wenty introduced Len to Muriel Bullen, a young lady from Brockhollands.

Just after the outbreak of war, on 28th October 1939, Len and Muriel married in Clifton Parish Church. Len was still working as a dental mechanic but was soon to change his job to that of fitter at the Bristol Aeroplane Company's factory at Filton. This was where he met his death. On the morning of 25th September, 27 German bombers, flying from bases at Dreux, Chartres and Villacoublay and escorted by about 30 Bf 110s, attacked the Filton Works from 11,000 feet. They killed 82 people and injured 170 others, most of the casualties resulting from high explosives striking trench shelters. The damage to the factory was severe.[(9)]

Len was one of the 82 fatalities resulting from that raid. His first wedding anniversary would have been in just over a month's time. He was aged 34.

Dillwyn Jenkyns
Jenkyns not Jenkins

Sam Dillwyn Jenkyns was born on Sunday, 9th November 1919. He was the second son of four children born to Thomas Aeron Jenkyns and Louisa Jenkyns. He was born in a cottage in what is now Blue Rock Crescent.

The family moved to 2 Parkend Road in the mid-1920s. Dillwyn, as he was always known, attended Bream School until the age of 11, when he won a scholarship to Lydney Grammar School. When he left school he became a trainee dispenser at Clifford Hughes Chemists, The Cross, Lydney.

In 1939 Dillwyn volunteered for the R.A.F. He became a navigator in Wellington bombers and flew missions over Kiel and in North Africa. He died in tragic circumstances, at Fayib in Egypt, on Friday, 29th November 1940. He was killed in a road accident involving a canvas-topped lorry. At the time he was a sergeant (No. 581458) serving with 70 Squadron.

When his name was carved on Bream Cenotaph it was mistakenly recorded as 'Jenkins'. When the stone was replaced by a bronze plaque, the error was rectified and he is now correctly recorded as D. Jenkyns. Dillwyn was 21 just three weeks before his tragic death.[10]

Dillwyn Jenkyns
Photo courtesy
Lloyd Jenkyns

Jack Neilson
'Jack Peacock'

Jack Neilson was one of the more difficult names on the Bream War Memorial to trace. Like Alfie Brooks in the First World War, Jack's name appears on two memorials and is spelt differently on both. His name is correctly spelt Neilson on the Bream Memorial but is incorrectly recorded as Nielson on the Tower Hill Memorial in London and on the Commonwealth War Graves list of Merchant Navy fatalities.

John Neilson was born on 20th June 1908 at Trowgreen. His father was Charles Neilson, a farm labourer. His mother was Esther 'Hettie' Neilson, formerly Blayden. In the early 1930s, Hettie Neilson, by then a widow, was living in Saundersgreen with her son. Jack, known by the intriguing nickname of 'Jack Peacock', went to Australia in 1932 on the '£10 Scheme' – a cheap one-way ticket. Eventually he decided to return to England but couldn't afford the fare. Consequently, he stowed away on board ship, was discovered and, on docking at Southampton, spent a number of weeks as a guest of His Majesty in Southampton Jail, before finally arriving in Bream. This has to be the longest trip home from Australia on record!

At the time of war Jack was in the Merchant Navy. Because of the operation of German U-boats in the Atlantic, the overall casualty rate in the British Merchant Navy during World War Two was higher than in any of the armed services. Yet they were all volunteers and at the end of a voyage there was no compulsion to sign on for another trip. Nor was it a well paid job; the rate of pay was £9 per month with 2s 6d (12½p) per day danger money. For five-and-a-half years these men kept Britain in the war. Had the battle of the Atlantic been lost Britain would have been forced to surrender for lack of food.

Jack's last voyage was as a fireman and trimmer on the *S.S. Kavak*. The ship, built in 1929, was part of the Moss Hutchinson Line and operated out of Liverpool. On 2nd December 1940 the convoy that *S.S. Kavak* was in was attacked in the North Atlantic.[11] The *Kavak* was torpedoed and sunk by a German submarine. The Commander, Captain Napier, and the crew of 24 went down with the ship. Their names are all commemorated on Panel 66 of the Tower Hill War Memorial. The dedicatory inscription reads as follows:

> *1939 - 45 The twenty-four thousand of the merchant navy and fishing fleets whose names are honoured on the walls of this garden gave their lives for their country and have no grave but the sea.*

Jack Neilson was 32.[12]

Dorothy Hancock
With or without an 's'

Comparable to the confusion that can arise over the surname Brooks or Brookes, is that of Hancock or Hancocks. There are two Hancock names on Bream War Memorial and both are incorrectly recorded as Hancocks. The mistake was made on the original stone list and was not amended when the bronze plaque replaced it. So for Dorothy Hancocks on the War Memorial, read Dorothy Hancock.

Dorothy Lilian Hancock was the daughter of John Cornelius Hancock (a miner) and Ellen Hancock of Bream Woodside. Dorothy was born in 1917 and had two brothers, Ernest and Douglas, and two sisters, Phyllis and Kathleen. She attended Bream School and was a member of Bream Girl Guides.

During the war, Dorothy was a munitions worker and, in common with others from the village, she made a daily journey by bus to work at the Royal Ordnance Factory in Hereford. Munitions work could be dangerous and it certainly killed Dorothy. She died in Lydney Hospital on 13th December 1940. Following a post mortem the cause of death was identified as *'Acute necrosis of the liver and kidneys due to T.N.T. poisoning'*.

The funeral was held at Parkend Road Methodist Church on 18th December and Dorothy was buried in Bream Churchyard. Her parents and her brother Douglas (who died in 1944) share the same grave. Wreaths at the funeral, apart from the family, came from *'all her workmates at Unit 4 R.O.F. Hereford'*, *'her friends on the 4015 bus'* and *'her friends on Willetts bus'*. Dorothy was 23.[13]

Bill Drew
'Loss of Control in Adverse Weather'

Bill Drew was the elder of two brothers. Both served in the R.A.F. and both died, Bill in 1941, Stan in 1944. Bill was born William Edward Drew on 7th May 1912 and baptised at Bream Church on 23rd June.

His parents, William and Mary Drew, had previously run The Miners' Rest on Bream Tufts but by the time Bill was born the pub had closed and become a private house. It is now owned by Pat Lewis.

Having been a publican, William Drew became a miner and tragically died at the age of 46 in 1915. At the time Bill was three and Stan was only nine months old. Both boys attended Bream School and then Lydney Grammar School. Bill was described as being clever at school, a classmate Ron Hewlett remembers him as *"always top of the class"*. He was also good at boxing and hockey.

In the 1930s Bill joined the R.A.F. He attended Cranwell and became a Pilot Officer (Service number 4356). In the R.A.F. he is believed to have met Aircraftsman T.E. Shaw. The name sounds anonymous – if so, it was intended

Bill Drew, on the right, photographed in September 1934 with some R.A.F. mates, identified only as, from the left, 'Snoggle', Jerry and Mick.
Photo courtesy Ann and John Palmer

THE DAILY MIRROR Monday, June 10, 1940

Bill

One of the bomber crews who played havoc behind the enemy lines bombing roads, troop concentrations, and blowing up ammunition dumps. They are just getting into their plane. Within an hour of the picture being taken the crew were well over enemy country.

Bill on operations (on the left), *Daily Mirror*, 10th June 1940. *Courtesy of Elsie Drew* to be! It was the name adopted by deed poll in 1927 by T.E. Lawrence ('Laurence of Arabia') when he sought anonymity from the public eye.

Bill married a girl from Barrow in Furness, Lancashire, called Edwardena, usually called Ena. At the beginning of the war he took part in the much ridiculed leaflet drops over German cities. The purpose was to turn the German people against Hitler but the practical impact was probably minimal, except in providing the population with toilet paper!

Bill served in heavy bombers. The photograph of Bill boarding a bomber in the *Daily Mirror* on 10th June was taken at a time when British morale was low, Dunkirk had just happened and France was on the verge of surrender.

R.A.F. Bomber Command were to lose 9,000 aircraft during the war and at a staggering cost, for of the 125,000 aircrew that served in the various squadrons, 55,000 were to become fatal casualties and a further 18,000 were wounded and/or taken prisoner. It is a common misconception that Bomber Command losses were all sustained over enemy territories. This is not the case. Some 15 per cent. of losses occurred in training and other accidents, many of them in the U.K. This was what happened to Bill. On 13th April 1941 he was flying as the observer on Avro Anson N9857 of No. 19 Operational Training Unit (OTU), based at R.A.F. Kinloss, Scotland. The aircraft had taken off from Kinloss to undertake a Cross Country Navigation Exercise. Aircraft and crew failed to return. The last point of contact was between Tarbat Ness and Stoer Point and, initially, it was thought that the aircraft was lost in this area. However it was found nearly six weeks later by a shepherd on Ben More Assynt, a remote mountain in Sutherland. The cause of the crash was not established but it was thought that it was possibly due to *'loss of control in adverse weather'*. There are pieces of the aircraft still on the mountain which is located at Map Reference 294232 OS Sheet No. 15.

Bill Drew's grave on Ben More Assynt, Sutherland. *Photo courtesy Elsie Drew*

Apart from Bill, there were five other members of the crew who died on that mountainside. Because of its remoteness, they were each buried in the traditional manner under a cairn of stones at the scene of the crash. They are commemorated on a memorial at Inchnadamph Old Churchyard, Sutherland. Bill was 28. His death prompted his younger brother, Stan, to volunteer for the R.A.F. Bill's morse key is still at the old house in The Tufts.(14)

The Blackest Day

For every generation there seems to be one defining event so momentous that people remember where they were and what they were doing for the rest of their lives. Recently one such event was the death of Princess Diana. On 22nd November 1963 it was the assassination of John F. Kennedy. On 24th May 1941 it was the sinking of the battle cruiser *Hood*.

If any one ship was the embodiment of British sea power between the wars it was the *Hood*. The Royal Navy had ruled the seas since – and even before – Trafalgar in 1805 and the mighty *Hood* was the personification of Rule Britannia. At the time she was the biggest warship ever built – 880 feet long, with a mighty armament of eight 15 inch guns and an unrivalled maximum speed of 32 knots. At top speed it took a ton of oil to drive her half a mile. Above all she was a beautiful ship with an elegance of line that drew admiration wherever she went.

In the inter-war years 'The Mighty *Hood*', as she was known, was a great ambassador for British Naval power, 'flying the flag' all over the world. With a quarter of the globe coloured red, it was inconceivable that the sun would ever set on the British Empire when it possessed such a ship as this. Her biographer, Ernle Bradford describes her as *'a legend'* which provoked awe wherever she went.

Yet the *Hood* was an old lady by 1939 and she had one serious flaw. She had been laid down before Jutland and, like the *"bloody ships"* that Beatty had cursed in 1916, she suffered from a lack of armour on her upper decks. All battleships built after Jutland had the fault rectified and it was always intended that the *Hood* would have a long refit to armour her main deck but with international crisis following international crisis in the 1930s it never got done. It was inconceivable to take the *Hood* out of commission for the several months the refit would take. The ship was always needed somewhere.

From the German viewpoint, the battle of the Atlantic was largely fought by U-boats, which were supplemented from time to time by surface raiders. In May 1941 the battleship *Bismarck*, the most powerful ship in the German Navy, broke out into the Atlantic. She had the firepower to wreak enormous havoc among the convoys if she were not contained. The *Hood* was sent to deal with the threat.

The duel between the *Hood* and the *Bismarck* seemed a clash of the titans but it was also a clash between a ship that had been laid down in 1916 and one that had been laid down in 1939. It was a mismatch. Six minutes after the battle was joined, a shell from the *Bismarck* pierced *Hood*'s weak armour plating on the upper deck and exploded in her magazine. A huge column of flame, 1,000 feet high, erupted from the *Hood*. Her back broken, the ship broke in two and with bow pointing towards the sky like a cathedral, the *Hood* slipped beneath the waves. In a last defiant gesture her forward turrets spat out a final salvo. In less than two minutes the mighty *Hood* had ceased to exist. Of her crew of 1,416 men only three survived.(15)

When the report came through to the Admiralty from the Atlantic it simply said, *'Hood has blown up'* and was marked 'secret'. When the news did get out, the reaction was one of disbelief and incredulity. Then despair. The nation was in shock. If *Hood* could not stop the *Bismarck*, what could? And then the wider question arose, subconsciously even if it wasn't voiced – what could possibly stop Germany winning the war? That day, 24th May 1941, was the blackest day for Britain in the war. The nation was traumatised. It was also the blackest day for the village. Two Bream boys were on the *Hood*, those very good friends Bob Carpenter and Doug Davies.

Bob Carpenter
'Our Bob'

Bob Carpenter was called 'Smiley' to the extent where it was believed to be his baptismal name. It wasn't, it was a nickname. Robert Stanley Carpenter was baptised in Bream Church on Friday, 5th December 1919. He was born on 22nd September 1919. His parents were William and Ellen (née Cresswick) Carpenter from Bream Woodside, where they lived next door to the Hancock family. William Carpenter was a miner. He died when Bob was still young, in 1924. Bob was one of a family of eight children; two sisters Elsie and Olive, and five brothers Osborne ('Burns'), Albert ('Bert'), William ('Bubs'), George ('Jim') and Harry (known as Harry!!).

After leaving school, Bob worked at Lydney Park Estate Gardens with his brother Harry. He joined the Marines (Number PO/X 3508) in 1938, took part in the Aldershot Tattoo and represented Devonport and Gloucestershire at rugby. He played in the front row.

Following the fall of France in June 1940, the French fleet posed a dilemma to the British Government. If it fell into German hands it would dramatically alter the balance of naval power. The French Admiral at Oran was given the options of scuttling his ships, joining the British fleet, or taking them to a British or French colonial port. He refused all of these. The Royal Navy then attacked their former allies. Anglo-French relations, already strained to the limit by Dunkirk, became very bitter. Fifteen hundred French sailors died. The *Hood* was part of this bombardment. It must have been a particularly difficult action for the crew – six months earlier they themselves had served under the self same French admiral, Gensoul, in a combined operation off Iceland.

On his final leave Bob told his brother Harry that he thought he would not return. His concern was over the *Hood*'s lack of armour. This view is corroborated by a story from Harold Watkins of Bream Woodside. Both the boys from the *Hood* were having a drink or two – or several! – on the last night of their leave at The Rising Sun. Harold invited them home for supper. Both told him they had a premonition they would not return.

The *Hood* sank on Saturday, 24th May 1941. Bob was 21 years old. At home his mother never locked the door of the house again stating that *"if anyone would get off the ship it would be 'Our Bob'."*

A family story about Bob relates that when he was asked by Dr. Pat O'Driscoll if he could kill someone, Bob replied, *"If one of us a gotta go then I shall see it yunt me!"* (16)

Left: Bob Carpenter.
Photo courtesy Ann Sargeant
Above: Doug Davies.
Photo courtesy Trevor Davies

Doug Davies

Douglas Alan Davies, born on 15th May 1917 in Whitecroft, was the second oldest of a family of five. His parents were Thomas and Bertha Davies. He was educated initially at Pillowell School, then at Bream School when the family moved to Brockhollands. He finished his education at the Craft School, Lydney (now Lydney C. of E. School). On leaving school he worked on a coal wagon until he joined the Navy at 16 in 1933. This was the year Hitler became Chancellor of Germany. Doug was very keen on the Navy and signed on for 12 years serving on warships including the *Nelson*, *Coventry* and *Resource*. He was drafted to the *Hood* on the outbreak of war and was present at Oran. He travelled widely in the Navy but always with a happy-go-lucky attitude to life. One day he was an able seaman, the next day promoted to leading seaman, the day after demoted back down to able seaman!

When his 12 years were up, Doug had planned to set up in business with his younger brother Trevor – either a poultry farm or a fish and chip shop. Unfortunately this was not to be. Doug had celebrated his 24th birthday just before the *Hood* left Scapa Flow on her last fateful voyage. He was also about to become engaged to be married to Doris Howard from Little Drybrook.

When a soldier or airman dies, the family at home do not usually know until the dreaded telegram arrives from the War Office or Air Ministry and the village Post Master or Mistress would be the first person in the community to know. When a capital ship like the *Hood* died it was impossible to keep the

Shipmates in life and death. Bob Carpenter, front left, and Doug Davies, front right, with two other members of the *Hood*'s crew. The photo is dated May 1941 and was taken just before the ship's last voyage. The seaman, back right, is believed to have gone to another posting before the *Hood* sailed.
Photo courtesy Gerald Vaughan

news out of the public forum and the sinking was announced on the radio late on Saturday, 24th May. In Bream Doug's father was at a concert singing with the Forest of Dean Male Voice Choir. Mum was at home but, by ironic coincidence, their radio had broken down for the first time. The first inkling that either parent had that something was badly wrong was when Dad got off the coach in Bream that night and was greeted with a sympathetic *"terrible news about the Hood"*. He then had to face the walk home and to break the news to his wife.

Apart from Bob and Doug, there are two other connections to the *Hood*. The ship was commanded by Captain Ralph Kerr C.B.E. (known, according to Doug, by the nickname 'Scarface Kerr' among the ship's crew), whose parents lived in Newnham and where Ralph Kerr usually spent his leave. A memorial service was held in Newnham Parish Church on Saturday, 31st May. Both Bream families were there to pay their last respects to their sons.(17)

Tommy Turley
"Always in the Middle of It"

Tommy Turley is unique on the Bream War Memorial. His name appears under his nickname and not under either of his chosen names. He was in fact Reginald Charles Turley but he was – and still is – universally known and remembered in Bream as Tom or Tommy. This is how he appears on the Cenotaph. I think he would be pleased by that.

Tommy was a 'character'. A contemporary of his at school said, laughing fondly at the memory, *"if anything went on at school Tommy was always in the middle of it."* (18) Similarly, when I spoke to Mary Robinson (now aged 100!) about Tommy her face lit up and she chuckled, *"He was an article"*, she said, *"full of mischief but no malice in him!"*

Tommy was a bit of a harum scarum character but loveable with it. Like man like boy, he retained the same capacity for being in the middle of the action when he grew up, and with a delightful irreverence when it came to authority. Like Doug Davies he was promoted in the Navy from able seaman to leading seaman but lost his stripe and was back where he started.

Tommy was born on 17th January 1911, the son of Louisa Turley who worked at the Rising Sun. He was born in Blakeney but grew up in Admiral's Lane (now Blue Rock Crescent), in a house called Parkhill View, with his mother, step-father Charles Wildin, and half-sister Ella and half-brother Ronald. Next door to Parkhill View lived Charles Wildin's brother, Thomas, and his family. Thomas served 22 years in the Royal Navy and he was followed into the service by his three sons, Walter, Leonard and Geoff.

On leaving Bream School at 14, Tommy worked at Ben Bath's Bakery for a short while but, probably inspired by his next door neighbour's naval stories,

Tommy Turley shows the King how to do it! Tommy (centre) is standing in front of King George V on the Royal Yacht. *Photo courtesy Geoff Wildin.*

he joined the Navy as a boy sailor in 1926, at around the same time as his step-cousin, Walter Wildin, whose name is also on the Cenotaph.

In the Navy he was an able seaman (D/JX 126754) and – temporarily! – leading seaman. One of the ships he served on from 1930-2 was the battleship *H.M.S. Queen Elizabeth*, flagship of the Mediterranean fleet.

Tommy Turley.
Photo courtesy Ella Dawe

Tommy grew to be six feet tall, and was a fit and athletic young man who regularly represented Devonport Naval Base in public displays against Chatham and Portsmouth bases. In the early thirties he was selected to help crew the *King George V* yacht and knew the King.

In 1937, while visiting New Zealand, Tommy met and married a girl called Christina Brown. They had a daughter called Tui May. This was probably the happiest time of his life. Heartache was to follow. In 1939 Tommy's ship, *H.M.S. Wellington*, returned to England as the crisis in Europe gathered pace. On his return Tommy received a letter telling him his wife was very ill. He could not get back to New Zealand. The next letter told him that Christina was dead and their daughter Tui May was being looked after by Chris' sister Tui. Sometime later Tommy had another letter saying that his daughter had also died, from meningitis. It broke Tommy's heart.

At the outbreak of war Tommy volunteered for the Naval Commandos. He was taken prisoner in May 1940 at Narvik and managed to escape and join a British Army group who were still free in Norway after the evacuation of the main body of British troops. He was then recaptured and imprisoned in occupied France. Again he escaped and again he was captured. Undeterred, Tommy escaped again – the third time! – and crossed from France to neutral Spain, where he reported to the British Embassy in Madrid. At this point, January 1941, Tommy's mother received a letter that he was safe in Madrid and awaiting repatriation through Gibraltar. She was advised not to go public on the information yet in case it prejudiced his chances of getting out of Spain.

The irrepressible Tommy did eventually arrive back in this country and was given three months' leave. After his leave, Acting Seaman Turley, a rank he held till his death, reported back to Devonport and joined *H.M.S. Neptune*, a battle cruiser, which was supposedly bound for New Zealand. He got this posting through the good offices of his step-brother Len, who was then in the drafting office at Plymouth. From the *Neptune* he wrote to his mother that he had had enough of the Navy and sailing the seven seas. He intended to become a civilian when he had finished his time. Probably he wanted to settle in New Zealand (hence the posting to *Neptune*) where he had been so happy just a couple of years before.

HMS Neptune – the ship on which Tommy died. *Photo courtesy Ella Dawe*

En route for New Zealand, the *Neptune* was on active service off the North African coast where it sunk a German supply ship bound for Rommel's North Afrika Corps. Soon after, on Friday, 19th December 1941, the *Neptune* hit a mine off the Libyan coast and sunk. Only one crewmember of the ship's complement of 766 survived. Tommy was 30. His luck had finally run out, although he would probably argue that it had run out two years before when he lost both his wife and daughter.

The final word on Tommy should remain with his step cousin Geoff Wildin, who remembers him for his devil-may-care attitude. Tommy had a capacity to get into scrapes but he always bounced back. The best Tommy story of them all is that, in 1941, on arriving home after his three wartime escapes, a grateful Admiralty decided that the hero should be decorated. Tommy, however, failed to turn up for the award ceremony. At the time he was in a police cell having been celebrating his escape a little over-enthusiastically the night before. That was Tommy! May he rest in peace.(19)

Tommy's name is commemorated at the Tripoli War Cemetery (Reference 5.E.19), the Naval Monument at Plymouth Hoe and on Bream Cenotaph. At Tripoli and Plymouth he is listed as R.C. Turley. At Bream he is correctly listed as T. Turley. 'T' for Tommy.

In the Village in 1940

In the village the early months of 1940 had seen many evacuees return home as the Phoney War continued. This trend was dramatically reversed with the evacuation from Dunkirk, which ended on 4th June 1940, and the impending fall of France. A bigger evacuation than that of September 1939 now took place. On 7th June 1940, 58 boys arrived in the Boys School, with their own teachers and, to preserve their own identity, were taught in two distinct classes. The boys, from Rochford, Essex, were finally merged into the Bream classes in January 1941. In the Infants School, 12 evacuees from Rochford and Sutton, Essex, arrived on 10th June with their teacher, Miss Baker. Early in July, fourteen hundred evacuees from Birmingham arrived in the Forest.(20)

Later in July, 60 mums from Rochford arrived by coach to visit the evacuees on a day trip.(21) Rather a long day for them.

It is difficult to appreciate how confusing it must have been for the evacuees. They had left their homes and, in some cases their parents, to go on a long journey to an area they did not know and live with people they had never met. The Rochford children left Essex at 8am on Sunday, 2nd June, travelled by train to Gloucester Station and then by coach to Bream. They arrived at midnight, were shepherded into the Miners Welfare Cinema and then chosen by the people who had volunteered to take them in.

In the early days, there were a few skirmishes and fights in the School playground – Bream versus Rochford – but that soon settled down and the evacuees were made to feel at home in the village.

Len Olley was one of the evacuees involved in the Boys School gardening programme. Three areas were used to 'Dig for Victory' and grow vegetables for the village. Firstly, the Boys School garden situated against the far wall from Parkend Road of the Community Centre. Secondly, by the searchlight battery billet, a hut by the entrance to Devil's Chapel (the searchlight battery itself was in a field belonging to Cook's farm on present day Greenacres). Thirdly, under the trees in the Girls School by the swimming pool. Len says the boys got into trouble over this, distracted by the girls doing P.E!

In September, Miss Baker was transferred to teach the Junior Girl evacuees and another 12 evacuees arrived in the Infants. In February 1941, evacuee numbers were such that the WVS started an evacuee club from 2pm - 5pm every weekday afternoon. In May 1941, 22 boys arrived from Filton and 15 more infants. Thereafter, as the Blitz diminished the number of evacuees reduced. From a high in November 1940 of 15 pupils, the Infants evacuee register dropped to five by July 1942. Bream Schools did not possess trenches or air raid shelters, thus, in accordance with the Ministry of Education Circular 352, on hearing an air raid warning the Schools were closed and children found shelter in a home five minutes away. Those who could not get home in the time allotted went to a friend's. One such entry in the School Log Book reads, *'27 Feb 1941. ARW at 3 pm. Boys sent home at once'*.

The Infants Log Book for 5th February 1943 records, *'The last official Evacuees left today for Southend'*. Some of course remained in Gloucestershire and even in the village. One such was Sylvia Innes, along with her elder sister Beryl and younger sister Jean. The girls' mother, Olive Innes, was formerly Miss Olive Watkins of Bream, and she had married a Scot, James Innes. Two of the girls, Beryl and Jean, were born in Bream but Sylvia had been born in London which was where the family lived. During the war, James Innes became a Sergeant Major in the Military Police and the rest of the family, bombed out twice in London, came to live in the Forest, initially at Yorkley and then at Bream, where both Sylvia and Jean were in the Infants School.

The Infants School Log Book for 18th September 1944 contains the following sad entry: *'Sylvia and Jean Innes have been transferred from Evacuee Register. Mother is a war widow and does not intend to return to London'*.

Sylvia thus officially became a 'Breamer' in September 1944. Her father, R.S.M. James Innes, had died of wounds in Italy on 3rd September 1943, aged 30. Sylvia has one fleeting memory of her father. He was crossing Parkend Road to catch the bus at the end of his leave.

Another who became a 'Breamer' was Mick Olley. When the evacuees returned to Essex in 1943 Mick hid in the woods, preferring to stay in Bream which he now regarded as home. He now lives abroad – in Alvington![22]

Fundraising and charitable war work grew in volume during the dark days of 1940 and 1941. A Sister Susie Club was formed in 1940 under the auspices of the Women's British Legion. The Chair was Mrs. D.K. Percival and they met in The Two Swans (a.k.a. The Double Ducks!).[23] Their purpose was easy to discover – knitting for the local servicemen and sending parcels overseas but their name intrigued me. Eventually I found the explanation in an old wartime song which is a real tongue twister. The chorus is:

Sister Susie's sewing shirts for soldiers
Such skill at sewing shirts our shy young sister Susie shows!
Some soldiers send epistles, say they'd sooner sleep in thistles
Than the saucy soft short shirts for soldiers sister Susie sews! [24]

I wonder if the ladies gave a rendition of this every time they met – similar to Liverpool fans at Anfield and *"You'll never Walk Alone"*! There are three accompanying verses and I have the music if anyone wants to attempt – publicly of course – what must have been a very considerable choral challenge.

The Welcome Home Fund was also very active, collecting money regularly and making presentations to soldiers home on leave – much the same as happened 25 years before. By October 1940, under the leadership of Major Percival, they had £40 in the fund and had made donations to 48 members of the Forces.[25]

The major fundraising effort in the village, the Forest and the country at large in the latter part of 1940 and the early months of 1941 was the Spitfire Fund. The Forest Fund was launched in October 1940. The target figure needed to build a Spitfire was £5,000. The Forest Colliery Owners initially gave £125 and pledged another £125 if the £5,000 was reached.[26] By 22nd November West Dean had raised £280, of which Bream had contributed only just over £2. George Peachey, the Secretary of the Bream Spitfire Fund, went public and wrote to *The Mercury* outlining Bream's plans:

'Although Bream is slow in making a start you can rest assured there is something moving.'

A house-to-house collection was due to start the next week; Bream Youth Movement were to arrange a mile of pennies, a School Concert led by Miss Burnham, a teacher from Rochford and Miss Drabble, Girls School Headmistress, was arranged; and the children were taking coppers to school.

The final figure raised by the village for a Spitfire was £104-18s-1d (£104.90).[27] The largest single contribution to that was from the School Concert, which took place on Thursday, 30th January 1941 in the Girls School. The performance consisted of two humorous sketches by Rochford boys and one by Rochford girls, an operetta by the Rochford girls and a play by the

Bream Girls, all interspersed with recitations and choral singing by Rochford and Bream Choirs. Mick Olley still remembers the performance and his solo rendition of *'The Bailiff's Daughter of Islington'*! The concert was a great success and amply illustrates the evacuees' enormous contribution to the village and how well the village took them to its heart.(28)

The Forest target of trying to build its own Spitfire had been hopelessly optimistic. The final figure raised was £2,795 8s 10d (£2,795.44), enough to cover the cost of an engine. The Fund was then amalgamated with the Chepstow and Wye Valley Funds and the plane had painted on its fuselage *'Monmouth, Chepstow and Forest of Dean Spitfire'*.(29) Its photograph appeared in the 15th August 1941 edition of *The Mercury* under the proud caption, *'This is Our Spitfire'*.

The Forest Spitfire initiative was followed by other local and national fundraising projects. A War Weapons Week was organised in 1941 and a Warship Week in 1942. The lessons of the Spitfire venture had however been learnt, the target for Warship Week was to be a submarine hull and not the complete works!(30)

As well as larger fundraising initiatives, there were smaller very local ones. In August 1940 the children of Parkend Road organised a show and raised 16 shillings (80p) for the men of the local searchlight section down Lydney Road. The organiser was Verley Clutterbuck.(31)

The end of the phoney war in May 1940 led to an appeal for a civilian force, aged 15-65, known as the L.D.V. (Local Defence Volunteers), to repel possible invasion. In the summer of 1940 the L.D.V. were renamed The Home Guard, a far more familiar name. Even more familiar is the Dad's Army label by which they were immortalised in the T.V. series of that name and are lovingly and endearingly remembered by so many.

Were they the Dad's Army of legend in Bream? Every community has its Dad's Amy stories and Bream and the Forest are no exception. A *Mercury* story on 12th July 1940 relates that several L.D.V. members (platoon unspecified) were in possession of rifles *'when not in full possession of their faculties'*, i.e. after spending an hour or two in the local pub!

When the L.D.V. were first formed in Bream, the C.O. was Major Percival of Priors Lodge. H.Q. was the skittle alley at the Rising Sun, which also served as an indoor rifle range (as it had in 1914!). H.Q. was also used for training. Like Captain Mainwaring, Major Percival was short sighted and when reading from the Training Manual to the company he had to have the book up in front of his face. As the lecture proceeded there was a gradual drift away (back row first!) into ‚Auntie Annie's,, i.e. The Rising Sun. A delightful picture!

Major Percival was also present at a training session to demonstrate the Browning Automatic Rifle. The rifleman was Maurice Jarrett. Unfortunately the Major lay down next to Maurice on the wrong side, i.e. the discharge side, and was hit in the cheek by the hot spent cartridges. He jumped up shouting *"I'm shot!"*. To be fair to Major Percival, I do not have his version of either of these anecdotes.

The equipment in the early days of the force was woefully inadequate. Stored in sacks in H.Q. were clubs – the purpose? – to knock out German parachutists when the invasion came! It is fortunate that the clubs were never issued!

On a more serious note, the Home Guard generally took its responsibilities seriously. The men of Bream, Yorkley, Parkend and Whitecroft formed 'A' Company, commanded by Howard Wintle, M.C., Headmaster of Yorkley School, and were part of the 4th Forest Battalion of the Home Guard. The Company was some 120 strong. The Bream Platoon of 'A' Coy were initially commanded by Major Percival, then Lt. William Watson the Head of the Boys School. Second Lt. Riley Lewis was second i.c., Dick Beach, Stan Worgan, George Peever, Don Warner and (for a short time) Evan Jones were sergeants. Many of the Bream Platoon were veterans. They were equipped with some Lee Enfield .303s and later, unused American P17 300s. They had a roadblock across the Coleford Road where the Winding Wheel was. They had an outdoor range at the Flour Mill Colliery (again exactly as in World War One!) and they regularly took part in Miniature Shooting matches at the Tufts miniature rifle range.(32)

On 17th August 1941 'A' Company fought each other! The Whitecroft and Yorkley platoons of 'A' Coy defended Whitecroft in a mock attack against Bream and Parkend platoons. The attack failed. The lesson learnt was that Whitecroft was an extremely difficult village for an attacking force to take. Anyone who has played sport at Whitecroft could have told them that!(33)

Collecting salvage was a major concern at this stage of the war. In Bream, the Girl Guides were collecting paper and the W.V.S. organised the collection of salvage. Virtually anything could be salvaged to reduce the need for imported materials and in 1940 an appeal was made for paper, bedsteads, pots and pans, old iron, brass, lead, aluminium, bottles, jars, tyres, rags and – wait for it – silk stockings! (32)

Provision for air raid protection was an issue between the village and West Dean Civil Defence Committee. Bream had three A.R.P. posts which were manned during air raids and the central one, which doubled as a first aid post, was manned for three and a half hours every night. It was situated at Bream School. Concern was expressed in December 1939 about inadequate heating and lighting. Further worries were expressed in August 1940, when Norman Fewings led a deputation of wardens to the West Dean Civil Defence Committee to voice their grievances.(35) Firstly, Bream needed its own siren to supplement the colliery hooter which was inefficient and couldn't be heard in the village unless the wind was in a certain direction. Secondly, the A.R.P. hut didn't have a toilet or table for the telephone. Thirdly, there were not enough wardens. Nearly a year later, in July 1941, the Civil Defence Committee were still discussing the lack of furniture in the post.[36] The Bream Wardens replied that the furniture was fit for the purpose and that they were more concerned with the fact that, even though it was a first aid post, there wasn't one blanket nor one stretcher in the place! When was Bream ever short of resources?

And how did the village react to this time of crisis in 1940 and 1941? Despite the threat of imminent invasion, Bream Flower Show was held at the Schools on Saturday, 10th August 1940. The Show was reported in *The Mercury* on 16th August under a defiant Churchillian headline:

BREAM SHOW GOES ON
NO GIVING IN AFTER 74 YEARS UNBROKEN RUN

Clearly Hitler was going to lose the war. He had failed to stop Bream Flower Show.

The Turn of the Tide
1942 - 1943

The year 1941 had seen the entry of Russia and the U.S.A into the war, and the next two years were to see the initiative switch to the allies. In January 1942, Sir Arthur Harris's appointment as C. in C. Bomber Command heralded an increase in R.A.F. operational sorties against Germany. August 1942 saw the Dieppe Raid, the first fighting British and Empire troops had done on the continent of Europe since 1940. The raid was a disaster but its failure provided valuable lessons for the invasion proper of Hitler's 'Fortress Europe' in 1944. In the Atlantic in 1942 the allies lost 1,700 ships but, by the middle of 1943, the British and the Americans had closed the 'air gap' in the central Atlantic and begun to win the battle of the convoys. The U-boats themselves became the quarry rather than the hunter.

On the Eastern Front, the German advance had been held up at Leningrad and Stalingrad. In January 1943, Von Paulus surrendered the remains of the German Sixth Army at Stalingrad. In July the Germans were again defeated, this time at Kursk in the greatest tank battle of the war. After Kursk, the Red Army began an offensive that was not going to stop until it reached Berlin.

In the Mediterranean, Montgomery's victory over Rommel at El Alamein in the autumn of 1942, as well as being the British Army's first success of the war, led to the surrender of German and Italian forces in North Africa in May 1943. The allied invasion of Sicily and Italy soon followed.

The turning point in the Far East came with the American naval victory at Midway in June 1942. From now on the Japanese were on the defensive against the Americans, although it was to be nearly another year before the British were to gain the upper hand in Burma.

Six Servicemen from Bream died in 1942 and 1943. Two were members of R.A.F. Bomber Command, one was a naval officer and three were soldiers. All three of the soldiers died as a result of accident or illness.

Percy Meek
'Shammy'

Percy, or to give him his full name, Percy Charles Meek was born at the top of Pig Street in Bream Woodside on 16th May 1911. Baptised at Bream Church on 22nd July, he was one of nine children born to Richard and Elizabeth Meek. Percy attended Bream School and was a member of the School football team that won the Inter-Schools Cup in 1924-25. 'Shammy', as he was always known, played inside left in the team, Ron Hewlett was outside left and the two were great friends. In those days, well before 4-4-2, 4-3-3 or 3-5-2, football teams played a 2-3-5 formation and the forwards – five of them – played as a 'W' shape. As a system of play, it lost popularity in the 1960s.

The photograph of Shammy reproduced here is the only one of him I could find and it is taken from the team picture of 1925.

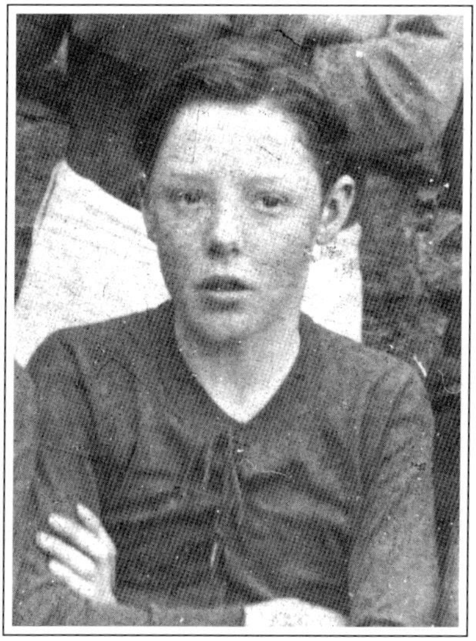

Percy Meek. Taken from the Bream School Football Team photo, 1925.
Photo courtesy Ron Ewlett

After he left school, Shammy worked at Princess Royal Colliery and continued to play soccer and rugby. Sporting prowess was in his genes, his elder brother Reg, known as 'Stodger', captained Salford Rugby League team! I'm not sure which is the better – or worse – nickname 'Shammy' or 'Stodger'!

Percy was in the war at an early stage. He was a gunner (No. 1778509) with 108 Battery, 134th H.A.A. 'Z' Regiment Royal Artillery. In 1942 he was taken seriously ill and died in Stanfield Sanatorium, Burslem, Staffordshire on 26th March 1942. The cause of death was tuberculosis.

Shammy was brought home and buried at St. James' Church on 31st March. Born, baptised and buried in Bream, he was 30 years old.[(1)]

Reg Moore
'Tempting fate?'

Reginald Moore, usually known as Reg or by the very distinctive nickname of 'Coot', was born in New Road, Bream, in the early months of 1916. He was the fourth of seven children born to Charles and Elizabeth Moore (née Jones). Reg was educated at Bream School and then Lydney Grammar School. His first job on leaving school was to work as a secretary at Lydney Grammar School for the Headmaster, Mr. Birch. He later moved to Camberwell, Surrey, to do a similar job at Wilson's Grammar School.

It was while at Camberwell that Reg joined the Royal Air Force Volunteer Reserve. He became a Sergeant Observer (No. 1164946) with 102 Squadron which by 1942 were mainly flying Halifaxes, one of the new breed of heavy bombers which had only come into service in 1941 and, with the Lancasters, were to form the mainstay of the great R.A.F. night offensive against Hitler's Europe.

Reg Moore. *Photo courtesy Hazel Sterry*

August of 1942 saw Bomber Command use P.P.F. for the first time in bombing operations. P.P.F. was Path Finder Force, which had the task of dropping flares ahead of the Main Force to assist in target identification. It was in one such raid against Dusseldorf on the night of 10th-11th September that Reg lost his life.

The plane that Reg was in took off from R.A.F. Pocklington in Yorkshire at 20:11 hours on the evening of Thursday, 10th September. It was a Halifax Mark II (No. W7752 DY-R) and had a crew of eight. It crashed into the North Sea. Six bodies were recovered from the water and taken for burial in various Dutch cemeteries.

Reg is buried in Terschelling (Westerschelling) Cemetery (Grave 72) on an island at the entrance to the Zuyder Zee, Holland. The date recorded on the grave is 11th September, implying the plane was on its way back from Dusseldorf. There are 55 other British airmen buried in the cemetery. From one point of view the raid was a success because it caused serious damage but the R.A.F. paid a heavy price for it. Thirty-eight bombers out of a total force of 479 failed to return, an unacceptably high loss rate of 7.9 per cent.

We all tempt fate at some time or other in our lives. On one occasion on leave in the village, Coot had been drinking in Aunt Annie's (The Rising Sun) with two of his pals, Darryl Camm, who was in the Navy, and Hylton Miles. The three emerged somewhat the worse for wear and both Darryl and Coot went across the road and chalked their names on the Cenotaph. Hylton called them both *"Bloody fools"*. Darryl then went back and rubbed his name out. Coot didn't. Darryl survived the war.

Reg 'Coot' Moore was only 26 when he died. He was due to marry a girl from Yorkshire, Eileen, in five days' time.

Coot's name is also commemorated on the R.A.F. Memorial in York Minster as well as on Bream Cenotaph.[2]

Reg Moore's grave at Westerschelling in Holland.
Photo courtesy Hazel Sterry

Sam Wilks
'a promising young man'

Sam Wilks had connections with two Forest communities and his death is consequently recorded on two Memorials, Bream and Christchurch, Berry Hill.

Samuel George Wilks was born at Yew Tree Cottage in Bream Eaves on 5th December 1916, the only son of Arthur and Ethel Wilks (formerly Miles). After he left school, he was employed as an engineer at Watts in Lydney. In 1939 Sam married Maisie (May) Manns, the daughter of Mr. & Mrs. Manns of 13, Coversham Villas, Berry Hill. He then moved to Berry Hill to live.

Called-up at the beginning of the war, Sam's natural inclination took him into the Royal Electrical and Mechanical Engineers (R.E.M.E.). Here he became a Corporal (No. 7621609).

For most of his service career Sam was billeted in the Shetland Islands and it was there that he caught the cold which led to tuberculosis. Life could be tough in the Army even without the complication of active service. He was treated at the Isolation Hospital at Over in Gloucester but died there on

Sunday, 27th December 1942. The funeral was held at Christchurch on Saturday, 2nd January 1943.[3]

Sam's headstone in Christchurch records the following epitaph:

'Thoughts go back
To happier days
Life goes on
But not the same.'

Sam was just 26 and left a widow and young son. His obituary in *The Mercury* described him as *'a promising young man'*.

Sam Wilks.
Photo courtesy Ronald Wilks

John Overington
'mission number 28'

Oliver John Overington was born in 1920 and lived at The Blistors in Bream. He was named Oliver after his father, Oliver Evans Overington, but was always known as John. His mother was Margaret Overington (formerly Davies).

John attended Bream School and then Lydney Grammar. He then worked for Imms the builders in Lydney.

Mad keen on flying, John signed up for the Royal Air Force Voluntary Reserve – the same route into the R.A.F. as Reg Moore. Like Reg, John trained for heavy bombers and served as both an air gunner and a wireless operator. He held the rank of Flight Sergeant (No. 1189925).

At this stage of the war, Bomber Command was regularly mounting night raids of upwards of 800 bombers against major German cities. John met his death on one such raid against Mönchengladbach on the night of 30th/31st August 1943. He was a wireless operator on a Lancaster Mark III (No. ED551 WS-M), part of 9 Squadron, 5 Group, based at Woodhall Spa in Lincolnshire.

With its crew of eight, the Lancaster took off from Woodhall at eight minutes past midnight. The plane was lost without trace. Thirty-two planes failed to return from that night's raid out of a total of over 800 aircraft taking part, a loss ratio of less than 4 per cent.

During 1943, Bomber Command lost 2,841 planes on operational sorties, the largest number of these (697) being Lancaster IIIs. 9 Squadron alone lost 43 – a very high casualty rate.

All of the crew are commemorated on the Runnymede Memorial. John's name is also on the Bream Cenotaph.

John had just been home on a 48-hour pass. He was looking forward to getting his next couple of missions out of the way because he would then have reached the magic figure of 30 ops. Thirty ops constituted a completed tour of duty and anyone reaching that figure was entitled to an extended break from operational flying. However, with the officially 'acceptable' loss rate on bombing raids of four to five per cent. not always mirrored by reality, a crew needed an element of luck to complete a tour. It was reckoned that crewmembers had a 3 in 10 chance of survival in 1943.(4)

The raid over Mönchengladbach was John's 28th mission. He was only 23 years old.(5)

John Overington.
Photo courtesy Reg Overington

Walter Wildin
'They Shall Not Grow Old'

Blue Rock Crescent in Bream was, at one time, known as Admiral's Lane. It was called this by virtue of the fact that a retired Admiral Jones originally lived at the end of it. Later, the naval connection became so strong that the place was virtually a naval base in its own right. Thomas Wildin lived here and he and his three sons, Walter, Leonard and Geoff, were all in the Navy. And if that were not enough, next door lived Tom Turley.

Walter Thomas Richard Wildin (usually known as Walter but 'Twinkletoes' in the Navy!) was born on 21st October 1909 at Devonport, the eldest of five children born to Thomas and Helen Wildin. Thomas moved his family to Bream during the First World War, to escape any possible German bombing raids on the naval base. Thus Walter was brought up in the village, attended Bream School and at 14 he left to work in the Pin Factory at Whitecroft. He was just marking time until he was old enough to join the Navy. He joined in 1925 as a boy sailor, just prior to his sixteenth birthday. He rose steadily through the ranks; October 1927 ordinary seaman; January 1929 able seaman; May 1936 leading seaman; December 1938 petty officer; and finally, 30th May 1940, gunnery officer. His service number was D/J 115161.

On 18th February 1939, Walter married Dora Essie Worgan, who was always known as Essie, at St. James' Church. At the time Walter was serving on *H.M.S. Victory*. The couple set up home at The Tufts.

During the war Walter was involved in a number of naval actions. He was at the battle of the River Plate in December 1939 and led a boarding party from *H.M.S. Diomede* on to the German freighter *Idarwald*. Scuppered by its own crew, the *Idarwald* subsequently sank. The German flagship, the battleship *Admiral Graf Spee*, was also sunk by its own crew in the River Plate and its commanding officer, Captain Langsdorff, committed suicide.

Grievously for Walter, his wife, Essie, died on 24th October 1941. She was just 31 and they had been married two-and-a-half years but with little opportunity to see each other.

Walter subsequently served on the Arctic convoys, running supplies to the Russian port of Murmansk. He was also involved in the landings at Salerno in Italy in September 1943. Six weeks later he met his death off the Channel Islands serving on *H.M.S. Charybdis*.

H.M.S. Charybdis was a light cruiser. As part of 'Operation Tunnel', it was leading a flotilla of six destroyers on the night of 23rd/24th October in a sweep west of the Channel Islands, beyond the range of the German coastal batteries there, hoping to surprise a German merchant ship out of Brest. It was the Royal Navy that got the surprise. The merchant ship was escorted by E-boats which had been warned of the Royal Navy's presence by coastal radar. The *Charybdis* spotted the E-boats on her radar at a range of 9,000 yards but was struck by two torpedoes before she could open fire. Out of a crew of 569, 30 officers and 432 ratings were lost. One destroyer, *H.M.S. Limbourne,* was also sunk with the loss of 42 lives.

23rd October 1943 saw the worst loss the Navy suffered in the Channel during the Second World War. The operation was severely criticised in the subsequent enquiry:

> *'A mixed force with no collective training, no knowledge of what enemy forces were at sea, and with deliberately-accepted disadvantages of light, was about to encounter a highly-trained striking force which almost certainly knew the numbers, position, course and speed . . . of our own force . . .* Charibdys *was more of a handicap than an asset in a night action in narrow waters . . .'*

Hindsight is a great teacher.

Fifty-one bodies were washed up on the Island's beaches, 21 in Guernsey, 29 in Jersey, 1 in Sark and over 100 in France. On Wednesday, 17th November, the bodies in Guernsey were buried at Foulon. The Germans wished to keep the event low key and the public were not invited but who can stop a grapevine? Five thousand islanders turned up with over seven hundred wreaths, one of them a huge anchor about nine feet across.

On Jersey, too, the islanders took the opportunity to make it plain that three years of occupation had not diminished their loyalty to Britain.

November 17th is still commemorated on Guernsey as Charybdis Day and there is an annual service held for those who lost their lives.

The Wildin family were very proud of Walter. He rose through the ranks and became an officer. He was proud too – even trying, jokingly, to get his step cousin and great friend Tom Turley to salute him when they were both on leave in the village. Tom's reaction is best left to the imagination!

Walter's R.N. badge and his medals are still held by his younger brother Geoff. As well as being listed on Bream Cenotaph, Walter's name is recorded on the Plymouth Naval Memorial, Panel 78, Column 1. His name is also carved on his wife's headstone in Bream Churchyard. The day the *Charibdys* sank would have been his wife's 33rd birthday. His own birthday, his 34th, had been two days before.

The epitaph on the grave reads:

'They shall not grow old.'

Walter was the only officer among Bream's World War Two dead. He had served 18 years in the Navy.[(6)]

Above: Eric Grindle.
Photo courtesy The Mercury

Left: Walter Wildin
Photo courtesy Geoff Wildin

Eric Grindle

Eric was one of those on the Cenotaph who was called by his second name. He was always known as Eric but his full name was Victor Eric Grindle. He was born in 1917, the son of Leslie and Elsie Grindle (née Wright) of Whitecroft Road. He was the eldest son in a family of six, his sisters being Marjorie, Doreen and Kate, and his two brothers being Mick and John. Before he went into the Army he was a painter.

As mentioned before, life in the Services carried risks other than active service. There was always a great risk of disease and illness allied to the dangers involved in training or 'exercises'. Eric died in training on New Year's Eve 1943 with a fractured skull, due to being crushed by the overturning of a Bren Carrier in which he was a passenger. The accident happened on Clayton Hill, Pyecombe, Sussex. Eric's wife lived in Kensington in London but Eric was buried in his home village, at St. James, on Thursday, 6th January 1944 (Row D, Grave 63).[7]

Eric had been in the Army for four years. He was a private (No. 5728828) in the 4th Battalion Dorsetshire Regiment. The unit was in training for the forthcoming invasion of Normandy. Eric was 26 years old.

The Village in 1942 - 1943

Funding the war effort continued to be a high priority nationally and locally in 1942 and 1943. 'Warship Week' took place in the week of 14th to 21st March 1942. The target for Lydney, West Dean and East Dean was to raise £175,000. Fund raising was led by Mr. John H. Watts, Chairman of the Forest Central Committee and people were encouraged to lend rather than give. The lending could be via Savings Bonds, National War Bonds, Defence Bonds, Savings Certificates, Post Office Savings Bank or Trustee Savings Bank. In the event, Warship Week exceeded all expectations in the district, realising £261,455.(8)

In 1943 the major national fundraising initiative was the R.A.F. 'Wings for Victory Week', to be held from 5th to12th June. Lydney, West Dean and East Dean were challenged to contribute £60,000 each. As part of its fundraising effort, Bream organised a Wings for Victory Fete on the Rugby Field on Thursday, 10th June. The programme included such delights as:

1. *A Home Guard Demonstration*
2. *Girls Folk Dancing*
3. *Mushie, The only Forest bred Wrestling Lion in Great Britain with his boyfriend Speck*
4. *Chasewater Charlie, a cycling clown act*
5. *Professor North's Punch and Judy*
6. *Sticking stamps on the Bream bomb*
7. *Parkend Silver Band*

The Fete was organised by Oliver and Margaret Ovington of The Blistors. After all they did have a son, John, serving as a wireless operator in Bomber Command. As previously described, John was to die less than three months later.(9)

Other fundraising in the village included flag days for the Soldiers, Sailors and Airmen's Families Association and Aid to Russia Fund.(10) On Saturday, 17th July 1943, the Red Cross Benevolent Section held a fete at the Rugby Field, owned at the time by Mrs. Jones, and raised £500 for the Red Cross.(11) In addition there were regular Red Cross collections in the village. The Girls School under Miss Drabble donated £10 to the Merchant Navy Comforts Service(12) and the West Dean National Savings Group used the local Schools as a means of saving. In the respective Infants, Girls and Boys Schools, 40,104 and 73 families were using the scheme. Every little helped – one young lad, *'Master David Rudge'*, even organised a concert in his garden and raised two shillings (10p) on a basis of *'a penny a head'*. I presume the penny was to get in!(13)

In October 1942, the Welcome Home Fund amended their rules by stating that all beneficiaries had to have resided in the district for at least two years.

They continued to collect regularly and hold monthly meetings at the New Inn. By the beginning of 1943 they had raised over £300 and paid out £220 to men and women in the services. This comprised 109 payments of 30s (£1.50), 27 of 25s (£1.25) and 23 of 20s (£1).[14]

The Sister Susie Club was valiantly trying to cope with a shortage of wool and early in 1943 they sent all the socks they had 'in hand' abroad, and supplemented them with First Aid outfits. The W.V.S. were actively mending socks as well as still working on salvage collection with the help of the Girl Guides.[15]

After the initial doubts about sporting activities at the beginning of the war, local matches had revived. For instance, in 1942-3 Bream Amateur A.F.C. won both the Dilke Hospital Cup and the Lydney Hospital Cup. In 1943-4 they fielded a team in the Lydney and District Hospital League Divisions 1 and 2.[16]

The Flower Show, on which the sun never sets, inevitably continued. The 1943 show – the 78th! – was advertised on 30th July with the following intriguing headline:

'The INVASION of Bream by all sensible people will be on Saturday 7th August.'

Among the attractions was – one to exercise the imagination – a Bun and Treacle Race on horseback! It was, incidentally, won by K. Price of Cinderford, with G. Moore (Bream) second and L. James (Bream) third.[17] How it was organised is a mystery to me.

Bream British Legion, which had been formed in 1921, was one of the oldest branches in the Forest. Its officials were President Major Percival, Vice Presidents Mr. B.W. Johnson and Mr. W. Watson, Chair Mr. H.J. Davies, Secretary Mr. G.F. Hughes and Treasurer Mr. R. Cox. As always the Legion was actively involved in – amongst other events – 'Flowering the Cenotaph', the Poppy Day Appeal and the Remembrance Sunday procession.[18]

'Flowering the Cenotaph' was regularly done by the Ladies Branch of the Legion. Even though the Legion no longer exists in Bream, the lovely tradition of 'flowering' is continued by Sylvia Parry (formerly the evacuee infant from London, Sylvia Innes) and her husband Gordon.

The procession of 1942 is worth focussing on, not least because we have a photograph of it and a description in *The Mercury*. The photograph was taken from outside the present Rugby Club. The parade consisted of Home Guard and Civil Defence Services. It mustered at the Schools under Lieutenant Watson and was led by Bream Silver Band under Lieutenant Evan Jones. Extra seating had to be provided in the Church where Jack Kent (seen in the procession) laid the Legion Wreath on the Altar. The Vicar, the Reverend A.M. Parr, took the service.[19]

These were increasingly anxious and sombre times in the village and the procession reflects that mood. As does the Poppy Day Appeal. In 1941 this raised £30, in 1942 £37 and in 1943 a staggering £54.[20] These were terrible times and there were many names to remember. At the time of the Poppy Day collection in November 1943, a further 15 names were due to be added to the Cenotaph, those who had died 1939-43. And the end was still not in sight.

Remembrance Sunday procession 1942, passing Bream Rugby Club. The standard bearers were Mrs Frewings and George Hughes, the wreath was carried by Jack Kent and the Musical Director of bream Band was Evan Jones.

Photo courtesy Clive Lewis

The Collapse of the Axis 1944 - 1945

By the beginning of 1944, Hitler's Fortress Europe was coming under direct attack. Mussolini had fallen from power in 1943 and Italy had sought peace – only for the country to be occupied by the German Army. The allied fight through Italy was to be a long haul. They were held up at Monte Cassino for three months but eventually resumed their advance and captured Rome in June 1944.

The same month saw the invasion of France. Codenamed 'Overlord'. D-Day was 6th June 1944 and took place on five major beaches in Normandy (from west to east, 'Utah', 'Omaha', 'Gold', 'Juno' and 'Sword'). It was a tough struggle but German resistance in Normandy was eventually broken at the battle of Falaise Gap in August 1944.

Despite the success of the allied advance in France, London underwent a final Blitz from the V.1 'doodlebug' flying bombs and the V.2 rockets that were launched from the still occupied Low Countries. However, by March 1945 the Western Allies had crossed the Rhine into Germany and the Russians were moving in on Berlin from the east. Hitler committed suicide on 30th April and between 2nd and 8th May the German Army finally surrendered unconditionally. V.E. (Victory in Europe) Day was officially celebrated on 8th May.

In the Far East, the British held the Japanese at the crucial battle of Kohima in 1944 and subsequently re-captured Burma. The Americans continued their advance across the Pacific towards Japan, taking Okinawa in June 1945. But Okinawa cost 12,000 American lives as the Japanese continued to resist *'to the last man'*. The American President, Harry Truman, decided to use the newly developed atomic bomb, firstly on Hiroshima on 6th August 1945 and later on Nagasaki on 9th August. Japan surrendered unconditionally on the 14th. V.J. (Victory over Japan) was celebrated on 15th August.

The cost of the war had been enormous. Globally there were fifteen million dead and twenty-five million wounded. Britain and the Commonwealth had suffered 400,000 military deaths, 90,000 civilian deaths and 475,000 wounded. The U.K. had spent £20.5 billion on the cost of the war and by 1944 had five million in its armed forces (compared with half a million in 1934).[1]

In 1944 and 1945, British forces had been in action more extensively than at any other time in the war.

The price for Bream was high. There were another nine fatalities, three in the liberation of France, Stan Drew, Des Hancock and Nobby Baker; two in the Far East, Cyril Marshall and Cyril Morgan; one as a prisoner of war, Verdun James; and three as a result of illness, Tommy Challenger, Des Wintle and Doreen Brooks.

Cyril Marshall
'A Gentleman's Gentleman'

Cyril Samuel Marshall was the youngest of three children. He was born in the house immediately above Downham's Bakery on 6th April 1914. His parents were Thomas Samuel and Ellen Maud Marshall, known as Thomas and Maud. Thomas worked in the pit and then became a gardener.

Having attended Bream School, Cyril went into Gentleman's service at Eastnor Castle, near Ledbury, Herefordshire. At Eastnor he fell in love with Eileen Charlton, who was employed as a lady's maid. The couple were married in Bream Church on 30th March 1940. Both were 25 years old.

Cyril was called-up early in the war and became a gunner (No. 1690774) with the 25th Battery 28 Light A.A. Regiment, Royal Artillery. His sister-in-law believes he also served as a batman to a Captain when he was in the Army. She went on to describe Cyril as *"a charming young man"*.

In the spring of 1944, the war in Burma was at a critical stage. Kohima is the capital city of Nagaland, some 220 kilometres from the Indo-Burmese border. It was here in April 1944 that the Japanese advance to India was halted. The siege of the small British and Indian garrison at Kohima began on 5th April, when the Japanese 31st Division cut the road to Imphal. The garrison were defending several hills around Kohima but as the siege tightened only one, Garrison Hill, remained in their hands. Some of the fiercest fighting of the Burma campaign took place in the garden of the Deputy Commissioner's bungalow, around the tennis court. Eventually a relieving force, the British 2nd Division, broke through on 18th April to raise the siege. The Japanese now went onto the back-foot in Burma and were never again to threaten India.

British troops in Burma were more likely to suffer from illness (usually malaria or dysentery) than they were from wounds. This is why Cyril was in Kohima at the time of the siege – he was sick. There was both a hospital and a convalescent unit at Kohima. The defence of Kohima was so desperate that sick and convalescent troops who could hold a rifle were used in the line.

Cyril died on Friday, 14th April, four days before the relief arrived. He was 30 years old.(2) He is buried in the Kohima War Cemetery (Ref. 4.J.5). Near the entrance is a memorial to the 2nd British Division. It is a large 'Naga stone', which the Nagas themselves use to remember their dead. The stone was erected by the Naga people and bears the inscription

'When you go home
Tell them of us and say
For their tomorrow
We gave our today.'

The only other Bream boy to die in the Far East was also called Cyril . . .

Cyril Morgan
'Two Tours'

Cyril Stanley Morgan was born in Bream on 7th November 1930. He was the only child of Harry James Morgan B.E.M. and Annie Morgan (née Parker). Cyril's mother died when he was a toddler, his father married again and Cyril was brought up by his Uncle and Aunt in Birmingham. However, he was a regular visitor to Bream.

When the war started he was working at the Raleigh bicycle factory in Solihull – his paybook intriguingly describes his occupation as *'cycle adjuster'*.

Cyril was in the R.A.F. Volunteer Reserve and joined up on 18th November 1940, just after his 20th birthday. When he was in the R.A.F., he alternated his leaves between Birmingham and Bream.

In the R.A.F. Cyril started as a wireless operator, then took a navigation course and became an observer. He completed two tours (i.e. 60 missions) in Lancasters over Germany before he was sent to the Far East.

By this stage of the war he held the rank of Flight Sergeant (No. 1210551)

Cyril Morgan
Photo courtesy David Morgan

and was serving with 177 Squadron. His luck ran out when he was shot down on 27th December 1943, attacking Japanese shore batteries on the Burma coast. He was flying as observer/navigator in a Bristol Beaufighter – a two-seater fighter and torpedo bomber/attack aircraft. The plane was hit, he and his pilot, Flight Sergeant Grant, baled out over the water and became prisoners of war when they reached land.

Initially they were seen as spies and harshly treated. Eventually they ended up in a p.o.w. camp called Rangoon Prison in Burma. Being a p.o.w. of the Japanese was, for British servicemen, five times more likely to end in death than being a p.o.w. of the Germans. The Japanese never signed the Geneva Convention regulating the conduct of war and the treatment of prisoners. Conditions in the camps were generally brutal. The Japanese took 193,000 allied service personnel prisoner and nearly one third of them died in captivity. Of this p.o.w. total, 50,000 were British and over 12,000 of them died. In many cases, those who survived were to carry the scars for the rest of their lives. The recent campaign for apology and compensation is testimony to how bitter the few living survivors of the camps still feel.

Cyril died of dysentery eight months after being captured. This was confirmed by the pilot of the Beaufighter who had shared Cyril's captivity and visited his family after the war. The Commonwealth War Graves Commission records his death as being on 9th September 1944 but the notification of his death received by the family from the Air Ministry gives the date as 7th September. Whichever date is true, there is no disputing that Cyril was recently 24 years old and had served nearly four of those years with the R.A.F.(3)

He has no known grave. As well as his name being on Bream War Memorial, he is also commemorated on the Singapore Memorial (Column 435).

Stan Drew
'The Nightingale Sang In My Heart'

The 6th June 1944 is one of the most memorable dates of the Second World War and for some in Bream it was one of the saddest. Usually called D-Day or – as it was referred to by German Field Marshal Erwin Rommel – *"the longest day"*, it was the beginning of the liberation of France. The landings in Normandy were a great success and by midnight over 100,000 allied troops were ashore. Over 10,000 however were casualties.

Stan Drew was one of the earliest D-Day casualties. To keep the German High Command guessing, the allies had mounted a series of bombing raids since 1st June between Calais and Le Havre, to create the impression that the landings would take place in the Pas de Calais. On the night of 5th/6th June the deception continued, with Royal Navy vessels towing barrage balloons up the English Channel to produce big ship echoes on enemy radar. 'Operation Fortitude' succeeded brilliantly in distracting German attention away from Normandy.[4] German troops in the Pas de Calais were put on a *'stand-to ready for action'* footing. In Normandy no such order was given.

Cornelius Stanley Drew, universally known as Stan, was born on 25th July 1914. He was the son of William and Mary Drew (usually called Polly) of Bream Tufts and was the younger brother of Bill. Stan was baptised in Bream Church on Sunday, 25th October 1914.

Like Bill, Stan went to the village school and then on to Lydney Grammar School. He then went on to St Paul's College, Cheltenham. At school he was particularly good friends with Reg 'Coot' Moore.

Stan was an excellent sportsman. He played rugby for Gloucestershire in 1933-4, also for Bream, the Metropolitan Police and Old Tiffinians. He played as a centre or on the wing. He also enjoyed long distance running.

In June 1935 Stan joined the Metropolitan Police and it was in London that he met his future wife, Elsie. Their first meeting was in Berkeley Square and when I – predictably – asked Elsie *"Did the nightingale sing?"* back came the loving reply, *"Yes . . . in my heart."*

Elsie and Stan married on 31st May 1941.

Stan was based at West End Central Police Station. He helped in the East End during the Blitz and when he was off-duty, helped Elsie fire watching.

Stan joined the R.A.F. when elder brother Bill died in 1941. He trained in Canada for a year and was then sent as a Sergeant (No. 136993) to 515 Squadron, based at Little Snoring in Norfolk. He was flying as a navigator in two-seater Mosquitoes. The Mosquito was a very versatile plane. It had a variety of roles as a highly successful night-fighter, reconnaissance aircraft and even bomber. One of its attractions was its speed, greatly aided by the fact that it was made mainly of wood. The wood used was very light, being 1/6" thick, three veneers bonded together. Many of these skins were made at Pine End Works in Lydney.[5]

Probably as part of Operation Fortitude, Stan and his pilot, Flight Lieutenant W.R. Butterfield, took off from Little Snoring for north east France on the night of 5th/6th June. Radio contact was lost at 3am over Lille. At the same time allied airborne troops began landing in Normandy.

Elsie was notified by telegram on 6th June that Stan was missing. She also received a letter from his Wing Commander F.F. Lambert:

> *'I have held your husband in the highest esteem since he joined the squadron a few months ago . . .'* and *'He was extremely popular with both his fellow Navigators, Pilots and with all the ground crew, and he will be sadly missed by all here.'*

The German authorities reported that Stan had been buried in Chievres cemetery, near Mons in south west Belgium and some 50 kilometres east of Lille. Unfortunately, because of later military action in the area and the re-arrangement and renumbering of graves, Stan has no known grave.

Stan was 29 when he died. He left a widow Elsie and a two year-old daughter Pauline. His son, called Bill after his brother, was born in September 1944, three and a half months after D-Day.

The name 'Bill' had previously been agreed by Stan and Elsie, otherwise Elsie would have called the baby 'Stan'.

As well as on Bream Cenotaph, Stan's name is commemorated in the Police Book of Remembrance in Westminster Abbey.

Stan Drew.
Photo courtesy Elsie Drew

Des Hancock
'So Near Yet So Far'

At the same time on 5th/6th June as Stan Drew was flying across the Channel to north east France, Des Hancock was part of the 'P.B.I.' (the 'poor bloody infantry') in a landing craft in a rough sea heading towards Normandy, as part of Operation Overlord. The unfortunate infantry felt the full severity of the weather in their landing craft and were obliged to make liberal use of those items of equipment which the War Office, callously frank, calls *'Bags, Vomit'*. Even worse for the unit Des was in, the 2nd Gloucesters, they were not in the first wave to go ashore and having arrived off the Normandy coast at first light they were not ordered on to the beach until 11am.

Desmond Osborne Charles Hancock was born at Bream Tufts on 19th June 1924 and grew up at 7 Parkend Road. Baptised at St. James on 5th July, he was the second of six children born to George and Gertrude Hancock. After he had left Bream School, Des had the chance, as every boy did, of going into the mines but chose instead to work as a labourer on the roads, because he could not bear the thought of being underground. He was called-up, aged 18, on 21st January 1943, when he was required to report to Colchester to begin basic training. Des then joined the 2nd Battalion the Gloucestershire Regiment as a private (No. 14519125).

The 2nd Glosters knew they were to become part of the invasion of France when they were mobilised for active service at Clacton early in 1944. On 3rd June, the Battalion were taken from a sealed camp at Brockenhurst to Lymington, where they embarked on three landing craft for Southampton. At Southampton the invasion, originally planned for 5th June, was postponed for 24 hours because of the bad weather. Eventually the three landing craft carrying the Glosters took their place in the huge convoy moving out of Southampton for France on the evening of 5th June. They formed part of the largest amphibious operation of all time.

The 2nd Glosters were in the 56th Independent Infantry Brigade, which was to go ashore on Gold Beach. Originally they were to land at 10am but this was delayed until 11am. Their first objective – and a pleasurable one, because they were fighting their way through welcoming crowds (the Germans having withdrawn) – was the liberation of Bayeux. Subsequently, the Battalion saw action at Tilly and Lingevres. By early July they were based at Parfouru l'Éclin. It was here that Des met his death on Sunday, 2nd July 1944.

After the war, Sergeant Ernest Bevan visited Des's family and told them the circumstances of his death. He had volunteered to go for the mail and had just got out of the dugout and was about to put his helmet on when a shell exploded nearby. The concussion he suffered caused heart failure.

Des is buried in the Hottot-Les-Bagues War Cemetery less than 20 kilometres south of Bayeux. He had been 20 less than two weeks before his death and

Des Hancock
Photo courtesy Iris Hoare

Des's grave is near Bayeux, in France.
Photo courtesy Iris Hoare

had been on active service for less than one month. He was the youngest in the village to die in World War Two.

Every Christmas after 1944 that Des's father George was alive, he refused to sit down to a family Christmas dinner. He always said there was *"one missing"*.

Des was homesick before he went to France and went AWOL before D-Day to come home and see his family one more time. He got as far as Gloucester Railway Station but was then caught by the Military Police and returned to his unit. So near and yet so far.[7]

On the Cenotaph he is recorded as Hancocks in both the stone and bronze versions. The correct spelling of his name is Hancock.

Nobby Baker
'Nobby, Son Of Dosser'

During June and July 1944, the allies were building up resources in the Normandy beachhead in preparation for a break out. The fighting was tough and casualties were high. The terrain which characterises much of the Normandy landscape and known as the *'bocage'* – small fields boxed in by

Nobby Baker.
Photo courtesy Iris Burford

Nobby's grave in France.
Photo courtesy Iris Burford

hedges and sunken lanes – made it difficult to advance against a determined enemy. By 27th June the Americans had captured Cherbourg, a vital first port in France, which speeded up supplies. By 13th August the allies had broken out and were beginning to close the net on German troops still in Normandy in the battle of the Falaise Gap. By 21st August the net was closed, a further 50,000 Germans were prisoners and Normandy was liberated.

Nobby Baker died on 13th August, just as the battle of Falaise Gap was beginning.

He was baptised Norman Bertie Baker but he was universally known as 'Nobby'. He was born 11th February 1918 and baptised in Bream Church on Friday, 29th March. He was the sixth of eight children born to John and Lilly Baker. His father, John Baker, who worked at the Princess Royal, was always known as Dosser Baker and was a famous rugby player. He was a scrum half who played – in order of importance! – for Bream, Gloucester, and South West Counties, and was reserve for England.

The family still fondly hold a postcard to Jack telling him of his selection for Gloucester versus Llanelli. The card is simply addressed to *'J. Baker, Footballer, Bream'*. Who needed postcodes or house names or numbers, everyone in the village knew Jack 'Dosser' Baker and was proud of his achievement. Some still remember him now.

Nobby, son of Dosser, attended Bream School and then appropriately went to work at Downham's Bakery (where else would a Baker work?!). He was called-up early in the war. His sister's recollection is that he was initially with the Royal Irish Fusiliers and then the Highland Light Infantry, based at Lerwick in Scotland. At first a private, he became a Sergeant in the 10th Battalion H.L.I. (No. 7044141). He also became an instructor in the Bren gun.

He is buried at St. Charles de Percy War Cemetery some 45 kilometres south-west of Caen. Some of his letters home are still with the family. He was 26 years old.[8]

Verdun James
'The Reluctant Miner'

Like a lot of other boys in the village, Verdun worked as a miner at the Princess Royal Colliery after he left Bream School. However, he eventually left the pit on health grounds and because he was no longer in a reserved occupation he was called-up in 1939. Captured at Dunkirk, he was sent to a labour camp in Poland where the Germans ironically put him to work as a miner. Having escaped from the pit he ended up in the pit. He died in Poland in 1944.

Verdun James.
Photo courtesy Dolly Milner

Verdun was born in Tump Cottage, Pastor's Hill on 22nd November 1918, two years after the First World War battle that gave him his name. He was the fourth son in a family of ten children, born to Charles and Jane James. His father was a collier. He was baptised in the village, at Bream Church on Tuesday, 7th January 1919.

Verdun was called-up (*'deemed to have enlisted'* in official jargon) on 16th October 1939. After basic training, he was posted as a private (No. 5185838) in the 5th Battalion Glosters. The 5th Battalion were a Territorial unit and had been mobilised on the day war broke out. They had arrived in France as part of the B.E.F. on 15th January 1940.

Before he went abroad, Verdun had a feeling that he would not return. He voiced this to his sister Dolly and to Alec Vaughan. Alec tried to jolly him along saying that if he (Alec) could get through the First World War he was sure Verdun could the Second![9]

Verdun was posted to the 5th Battalion on 23rd April 1940 and could not have arrived at a worse time. Blitzkrieg was about to be unleashed on France, Belgium and Holland, and the 5th Battalion would be fighting desperately to survive as the British Army retreated to Dunkirk.

The 5th formed part of the Dunkirk perimeter at the tiny village of Ledringhem. On the 26th, 27th and 28th May, they came under sustained heavy attack. Surrounded by superior numbers of enemy, they held the village against all odds for 60 hours. Ordered to withdraw to Dunkirk as best they could, only 14 officers and 130 men, many of them wounded, made it out of Ledringhem. Verdun was one who didn't. He had been taken prisoner.

On its return to England, the 5th Battalion was reformed as a fighting unit and was in Normandy in 1944. For Verdun though, the war was virtually over. As a p.o.w. he was made to walk to Poland and was then held at Stalag VIIIB Camp (after 1943 known as Stalag 344). He was then moved to a village called Yaworzno, where he was billeted in the local school and made to work in the local mine. His letters home say that he was well treated but Verdun continued to do his little bit for the war effort. One of the officers who had

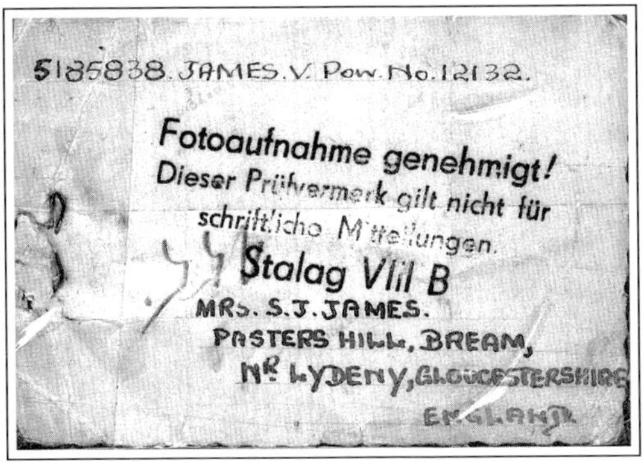

Right: Verdun's original grave at Yaworzno, in Poland.

Above: Reverse of the photo of the grave, sent from the camp, Stalag VIIIB, to Verdun's mother.

Photo courtesy Edna Pring

Verdun's present grave in Cracow.
Photo courtesy Edna Pring

been a p.o.w. with Verdun told Mr. and Mrs. James after the war that if Verdun thought the war was going badly for the allies he was not above carrying out a little underground sabotage!

In 1944 Verdun had appendicitis and before surgery could be carried out, peritonitis set in and he died. His military service records states that he died at Knappshaft Hospital, Hindenburg, on 15th July 1944. He was buried with full military honours in the village cemetery at Yaworzno. The grave was later relocated to the Commonwealth Plot in Cracow Rakowicki Cemetery (Ref. 2A.C.1), east of Cracow in Southern Poland. Verdun had been in captivity for four years. Within a few months of his death Poland had been overrun by the Red Army. He was aged 25.[10]

Tommy Challenger

Tommy Challenger died of illness after the war in Europe was over. He was born Thomas Henry James Challenger in May 1923, the only son of Thomas and Edith Challenger. The family lived at Oaklea, Bream Woodside. Thomas, the father, was an engine driver and died in November 1933. Edith, the mother, was the post lady and died in May 1967.

Having attended Bream School, Tommy worked for the Red & White Bus Company in Lydney. He was called-up during the war and joined the Royal Engineers as a crane driver, with the rank of private. He was serving in North Africa but was invalided home due to illness. The family recollection is that he was

Tommy Challenger.
Photo courtesy Dennis Challenger

suffering from fluid on the lungs. By this stage he was in receipt of an Army Disability Pension.

Initially he was in hospital in Hampshire but was later moved to Gloucestershire. He died in Gloucester Royal Infirmary on 27th July 1945, aged 22 years. His death certificate gives causes of death as a) meningitis, b) unknown. Sadly the problem with his lungs was never properly diagnosed but the family believed it was the result of the rigours of military training and active service.[11]

Tommy was buried at St. James Church on 31st July 1945.

Den Wintle
'Till We Meet Again'

Dennis, or Den as he was always known, was born in Brockhollands in 1915. He was the youngest of six children of Ben and Emily Wintle (née Rowland). Because he grew up in Brockhollands, Den attended Pillowell School. He then went on to work at the Tinplate Works in Lydney.

When he was called-up Den went into the Royal Navy. From there he was attached to the Merchant Navy on DEMS, an acronym for Defensively Equipped Merchant Ships. The DEMS programme was started in 1939, to provide protection for merchant ships against surface attack by U-boats by equipping them with guns and R.N. gun crews. The programme was successful in that the capability of U-boats to attack on the surface was much reduced and they were forced to attack from under the surface by torpedo. By March 1941, a total of over 4,000 British and Allied merchant ships were armed.

Den's naval career was largely spent in convoy work. His record of his war at sea, in particular his harrowing accounts of convoy runs to the beleaguered Mediterranean island of Malta, have survived in diary form. Unfortunately the diary is temporarily misplaced.

Den was taken ill on active service and invalided out of the Navy. He died at home of cancer on Friday, 30th November 1945. He is buried in Bream Churchyard.[12] His headstone reads:

D. WINTLE
Able Seaman RN DJX 314928
H.M.S. 'PRESIDENT III'
30th November 1945 Aged 30
TILL WE MEET AGAIN

Den died after the war was over. But there was to be yet another war death. Yet another name to be put on Bream Cenotaph. But that wasn't to be for another 12 months . . .

Den Wintle.
Photo courtesy Ken Kear

Dora Brooks.
Photo courtesy Maureen Williams

Dora Brooks

Doreen Evelyn Brooks was always known as Dora. She was step-sister to Alfie Brookes, who is on the First World War Roll of Honour. They shared the same father, Alfred Edwin Brook(e)s, who was usually known as Fred. After his first wife died, Fred married again in 1908, to Sarah Ann Lewis in St. James Church. He had three children by his first marriage, Alfie, Stan and Geoff, and there were to be another six by the second. Dora had two brothers, Percy and Fred, and three sisters, Beatrice, Ella and Myrtle. With the second marriage the 'e' seems to drop from the surname, 'Brookes' now became 'Brooks' but that may be purely coincidental.

Dora was born in Little Drybrook on 26th August 1920. Her father at the time was an 'onsetter' at the pit, the 'Bottom Colliery', i.e. the Flour Mill. After Fred Brooks died in April 1934, his widow and children moved from Little Drybrook to a house on Horwell Hill. The house has since been demolished.

Early in the war, Dora volunteered for the W.A.A.F, the Women's Auxiliary Air Force. This force had been created two months before the outbreak of war. Broadly its purpose was to use women in a supporting role, in suitable duties where they would release men for other work. Most were volunteers and they came to constitute more than 15 per cent. of the personnel strength of the R.A.F. They were paid a third less than most equivalent male ranks and received a smaller food ration. After joining the force, Dora was only at home when she was on leave.

Dora died of tuberculosis in Standish Hospital on 3rd June 1946 and there was certainly a feeling in the family that her wartime service had been a factor in her final illness. The funeral was at St. James Church on 14th June.(13)

Dora was 25 years old. The same age Alf had been.

And so the wheel had come full circle . . .

What began in the Brookes family when Alf died in France in 1915, ended in the same family in 1946 with the death of Dora in Standish Hospital, 31 years and 46 lives later.

Let us hope that the circle is finally complete and will never be broken by the addition of further names.

In the Village in 1944 - 1946

In 1944, fundraising continued to be of prime importance both nationally and locally. Salute the Soldier Week was designated for 3rd - 10th June 1944 and coincided with the closely guarded secret of the D-Day landings! The target was £75,000 from each area – Lydney, West Dean and East Dean – a total of £225,000. People were again invited to put money into Savings Bonds, National War Bonds, Defence Bonds, Savings Certificates, Savings Stamps, the Post Office Savings Bank and the Trustee Savings Bank. In June, the President of the War Savings Movement, Lord Bledisloe, announced that the target figure had been massively exceeded – £356,914 had been raised! (14)

In the village, the Welcome Home Fund continued its regular collections and its fundraising activities, such as competitions and dances. A key figure in all of this was the secretary, Mrs. E. Watkins. The fund also received donations from village organisations, including the Women's Service Club, the Methodist Sisterhood, the Special Police, the Royal Antediluvian Order of Buffaloes, Pine End and the Sport Clubs.(15)

Although by 1944 the Axis powers were in retreat and it was evident that the war would eventually end, there were still some scares on the home front. The V1 and V2 attacks in the South East led to requests in the Forest for host families to help with evacuees.

The impact of the war on local life could still be tragic. A U.S. lorry driver was found to have been *'criminally negligent'* when he killed 78 year old Emma Williams of Aylburton Common on Platt's Hill. However by then he was already in Normandy.(16)

Gradually though, the end of 1944 and early 1945 saw some return to normality in life in the village. On Wednesday, 1st November 1944, the Home Guard, though not yet officially disbanded, were *'stood down'*. On Sunday, 3rd December, the 4th Glosters (Forest) Battalion of the Home Guard held their final parade in Cinderford, finishing at The Triangle. One of the last public events that Bream Home Guard had taken part in was a grenade throwing competition at Bream Flower Show on 12th August. The entry fee for the competition was 1/- (5p) and the three prizes were £1, 10/- (50p) and 5/- (25p).[17]

In November 1944, complaints were received by the West Glos Power Company that they had not yet restarted street lights in West Dean, though they were on in Coleford. Bream, ever the poor relation! [18]

Some of the winding down activities at the end of the war had to wait for the fullness of time. The blackout curtains installed in School were not taken down from the Infants School until August 1945. They had been fitted 'late', on 24th August 1942, and came down 'late', on 27th August 1945. Then, as now, premises work in school has to wait until the summer holiday! [19]

Similarly, though the Home Guard had marched into history at their stand down parade in December 1944, the Special Constables did not hold their

Bream Sub-Section, Forest Police 1945. Back: Special Constables (SCs) F. Morse, W. Ellway, J. Windsor, H. Haddock, A. Parr, J. Reader, W. Robinson. Middle: SCs W. Lewis, G. Cooper, G. Meek, B. Lewis, A. Edmunds, W. Hill, J. Hancocks. Front: SCs H. Horsley, A. Thomas, Higher Special Constable W. Burdess, Police Constables (PCs) Weekes and Brown, SCs C. James, E. Edmunds. *Photo courtesy Graham Sacker*

final parade on Lydney Grammar School field until Sunday, 15th July 1945.[20]

Air Raid Precaution (A.R.P.) statistics reveal how lightly the Forest escaped during the Blitz. Of the four thousand high explosive bombs and twenty thousand incendiaries that fell on Gloucestershire between July 1940 and April 1943, only 220 and 620 respectively, fell on the Forest. There were 222 fatal casualties including six children in the county but none in the Forest itself, though it must be pointed out that the blackout in the Forest did cause fatalities – a possible case of the remedy being worse that the disease! The livestock count was 20 sheep and 20 cows killed in Gloucestershire. The only bomb to explode in the parish of Bream was at Brockhollands on 30th August 1940, on the same day that a cow was killed at Longhope.[21] Pat Lewis however tells me that seven bombs were dropped between the Bearse Wood and the road below Aylburton Lodge but none of them went off. The communal air raid shelter – the tunnel by the Flour Mill – was used less and less as the war went on.

When Victory in Europe finally arrived on 8th May, the headline in *The Mercury* was *'Flags, Fires, Dancing and Prayer'*. There was no 'mafficking', as there had been in 1900 during the Boer War. In Cinderford an effigy of Hitler was dragged up the High Street and burnt.[22] In Bream celebrations were a little more restrained. Children had Tuesday and Wednesday, 8th and 9th May, off school as a holiday. On Friday afternoon all three Schools had a Victory Tea Party outside under the trees. A staggering eight hundred cakes were made for this event. On 17th May the Welfare Hall Committee and the British Legion gave all the children a free cinema show, two tickets for free rides at the fair and 6d ($2^1/_2$p) to spend. On 3rd September after V.J. Day, the Bream Victory Committee held a Young People's Social in the School. The following day the School reported to the Committee that two notes on the piano had gone – even the greatest of celebrations could end on a sour note – or two! [23]

Towards the end of 1945 and early 1946, as service personnel were 'demobbed' or returned to the village after being prisoners of war, the village was busy celebrating the return of its sons and daughters. One such event occurred on Saturday, 28th October 1945, organised by Parkend Road in honour of Harry Preest, who had been a Japanese p.o.w. for three-and-a-half years, and Ken Watkins, who had been a German p.o.w. for a similar period. There was a carnival headed by the band, followed by a tea for children and old age pensioners. Then there was a concert in a marquee and a presentation to each soldier of an electric clock.[24]

Saturday, 8th June 1946 was officially to be the day of Victory Celebrations, 'V-Day'. By then everybody who was coming home was home. However *'the best laid schemes o' mice and men gang aft a-gley'*, particularly when it comes to the British weather. On 19th July 1919 all Peace Day celebrations in the village were cancelled because of rain – except for the bonfire on Sun Tump. History repeated itself with similarly bad weather on V-Day in 1946.[25] The programme for the day included sports events, a carnival for children, horse racing in Bream Avenue and a torch-lit procession at night, concluding again with a bonfire on Sun Tump. The carnival was cancelled because of rain but the sports were held under the able supervision of Messrs. N. Treherne and Ron James. However, the bonfire on Sun Tump was lit, as it had been in 1919, the only difference being that this time it lit up the Cenotaph, not yet built in 1919, which served as a reminder to the huge crowd present of what the two World Wars had meant to the parish in human terms. All the more poignant on 8th June because five days earlier had occurred the last of the parish's fatalities – Dora Brooks.

Bream Cenotaph

War Memorials have a very long history and ever since classical times the victorious dead have been commemorated in some form of monument. The 192 Greeks who died at Marathon in 490 B.C., defending the country against Persian invasion, were buried in a mound and their names were recorded in stone. As a War Memorial, the mound at Marathon was a celebration of victory; modern war memorials, however, are less about glory and more about the sacrifice and tragedy of war.

In the more recent past in Britain, monumental memorials have usually been statues of the heroic leader who was present on the field of battle – Henry V at Agincourt, Nelson at Trafalgar, Wellington at Waterloo. However, as war became more scientific and technologically advanced, generals retreated from the battlefield. More and more the battlefield hero, commemorated on war memorials, became the ordinary soldier.

This trend in Britain was apparent in the mid-19th century, when a few War Memorials appeared commemorating the Crimean War of 1854-6. The trend continued and the Boer War of 1899-1902 saw the erection of some 900 memorials in this country. This mood, prevalent at the time, of commemorating the ordinary soldier is well captured in Rudyard Kipling's contemporary poem *'Tommy'*:

"It's Tommy this, an' Tommy that",
an' "Chuck him out, the brute!"
But it's "Saviour of 'is country" when
The guns begin to shoot.'

The enormous loss of life in the First World War saw this trend of commemorating 'Tommy Atkins' reach a climax. The country experienced an unprecedented outpouring of grief, which led to the creation of some 40,000 War Memorials that involved every community in the land.

The most famous War Memorial in the country is the Whitehall Cenotaph. Initially, this was a temporary structure of wood and plaster, built in haste for the peace celebrations in July 1919 after the Treaty of Versailles had been signed. The architect was Edward Lutyens. The design was an empty tomb on a stepped pedestal, which served as an emotive symbol of all those service personnel who had died on 'a foreign field'. It immediately struck a national cord and was much admired. The temporary structure was replaced by a permanent stone memorial, unveiled on 11th November 1920.

Lutyens was adamant that he did not want a cross on the Cenotaph, he wanted it to be non-denominational. The monument was stark simplicity itself, its only decoration being a stone wreath and the inscription *'The Glorious Dead'*. Lutyens had originally wanted to use stone flags to decorate the monument but this was later changed to allow for real flags. The monument became nationally popular and was to inspire many copies and variations. Bream Cenotaph was one such copy, the only notable differences between itself and its Whitehall inspiration being that it is decorated with a cross and the inscription reads *'To Our Glorious Dead'*, not simply *'The Glorious Dead'*.

There was no centralised guidance from Westminster as to the form War Memorials should take. Thus the 40,000 monuments are considerably varied. Each community decided on the form of its own memorial, whether it would be a functional memorial hall, chapel or hospital, or a more decorative cross or cenotaph, or a simple commemorative tablet.

Similarly, the precise detail that was to be recorded about the fallen was also decided locally. In this respect also memorials vary enormously. Any permutation of the following detail can be found; surname, service, unit, service number, rank, gallantry awards, date of death, theatre of war, age and personal name in full or merely by initial.(1)

The names on Bream War Memorial provide the minimum of detail, simply a list usually with just one initial. Tracing them has been quite a historical challenge. Nevertheless, 45 of the list have now been traced, the only one who hasn't is T. Jones on the First World War list. Several possibilities have been suggested. Cyril Miles thinks he may have been a son of Vernon Jones the butcher. Hylton Miles has an idea he may have been from Clements' End Green. Recently, Margaret Addis has suggested he was a son of Arthur Jones. The fact is, from the village perspective, he is untraceable. Perhaps the 1901 Census when it is published next year, will throw some light on the matter. I certainly hope so.

If trying to trace T. Jones locally is difficult, it is well nigh impossible from the perspective of the military record. Statistically, the Commonwealth War Graves Commission has records of 7,825 Jones's who died in World War One, of whom 571 are listed as T. Jones, and, of course, there is no guarantee that T. was his only initial nor even his first initial. T. Jones is therefore my unknown soldier. I like to think of him as 'Tommy'.

The Cenotaph appeared on a number of picture postcards. This version was taken by Cinderford photographer R.G. Gibb around 1930, after the railings had been erected round it. *Photo courtesy The Archive Shop*

Another postcard view of the Cenotaph, in the 1930s.
Photo courtesy The Archive Shop

The First World War

Of the First World War list of twenty-one names, four appear with two initials, the other seventeen with one initial only. The single initial recorded for two of them is in fact their second initial, i.e. presumably the name by which they were known by their family and by the village. Four of the surnames have a different spelling to that recorded on their birth certificate. One name, identically spelt on both birth certificate and Cenotaph, is spelt otherwise in Army records. The list is alphabetical except that Mayne comes between Lucas and Matty but this, as suggested earlier, may be simply a stonemason error. Leslie Main is the spelling of the name on the birth certificate and on Army records. If he were to be listed as 'Main' on the Cenotaph, then the anomaly in alphabetical listing would disappear.

Of the twenty traced names, fifteen fell on the Western Front, three in Salonika, one in Gallipoli and one in Mesopotamia. Eleven of them were in The Glosters, which lost over 8,000 men in the war. All twenty were in the Army and none in the Navy or recently formed Air Force.

The style used in recording the names is very appealing. The fact that some names are listed under a second, rather than a first, initial and the variation of spelling in some surnames adds to the charm of the memorial. In a close knit community such as Bream, everyone knew who these men were and it was right and proper that they should be listed as they were known in the community.

A further appealing feature of the name recording is its simple democratic style. I personally approve of the fact that the monument makes no attempt to single out the officers or to indicate any winners of gallantry awards. In the democracy of death all names are equally listed because all made the supreme sacrifice.

The Second World War

The Second World War list contains more names than the 1914-18 list, 25 as against 21. This in itself is surprising because nationally the death toll was much higher in 1914-18.

Alphabetically, the 1939-45 list is significantly different. Names are listed alphabetically by the first letter of the surname and then chronologically according to year of death. Thus M. Bath is before N. Baker because Morris Bath died in 1940 and Nobby Baker in 1944. The listing is beautifully simple – all 23 men have one initial, the name by which they were locally known. This is carried to a logical conclusion where one initial is actually a nickname and not a Christian name at all! In a total of five names, the Cenotaph initial is not the first Christian name but the second or – in one case – the third!

The 1939-45 list is very unusual in the way it records the names of the two women who died. Nationally, when women were recorded on 1914-18 memorials, they were usually at the end, sometimes at the beginning and only occasionally alphabetically in a listing with men. However, on 1939-45 memorials the names of women who died are usually listed with the men in a non-discriminatory alphabetical order. At Bream this is not the case. The names of the two women are listed at the bottom but with their Christian names. Perhaps the discrimination balances out and can be said to be just, the men came first but are recorded by initial and surname, the women came last but are recorded by first name and surname. This way of recording is a fascinating social comment and is nearly unique to Bream. To my knowledge it only appears on one other memorial, Mildenhall in Suffolk.

As with the 1914-18 list there are anomalies over surnames. On the original stone Dillwyn Jenkyns was recorded as D. Jenkins. This has now been corrected on the bronze plaque but both Des Hancock and Doreen Hancock appear as Hancocks in both the stone and bronze versions of the list.

The 1939-45 list is varied in terms of the services represented. Nine were in the Army (only two this time in The Glosters), seven were in the R.A.F., five in the Navy, two were civilians, one was in the W.A.A.F. and one in the Merchant Navy.

Bream Cenotaph has a truly tragic tale to tell. Some families suffered more than once; two brothers died, as did a half-brother and sister, and an uncle and nephew. The average age of the people listed is 27. The oldest was 48 and the youngest barely 19. And how ironic that less than half of the names on the 1914-18 list even had the right to vote. The Franchise Act of 1884 only gave the vote to male householders. The Representation of the People Act 1918 extended the franchise to all males over 21 and females over 30 but that was too late for our soldiers. The feeling of tragedy is compounded by the fact that at least 17 children, some unborn, lost their fathers.

Several of Bream's names appear on other local War Memorials. Percy Frowen, Cyril Morgan and George Winn are on the Coleford Role of Honour. Ted Kilby and Sam Wilks are on the Christchurch Memorial in Berry Hill. Angus Dodgshon is also on the Lydney Cross. And yet there are some with a local connection whose names do not appear on any local monument. One such is Hugh Royal, who was born in Whitecroft and spent a lot of time in Bream before going into the Navy. Then there is Reg Batten who was actually living in Bream at the time of the First World War. Reg was a blacksmith who served in the Army. He died on 23rd December 1921. The stated causes of death were a) acute rheumatism (five months), and b) inflammation of the kidneys and haemorrhage. His sister, Carrie Davies, was convinced that the latter was caused by the fact that Reg had been badly gassed in the war.[(2)] Unfortunately Reg's name does not appear on the Cenotaph. By December 1921, when Reg died, the monument had already been built and the War

Remembrance Sunday Parade circa 1950, passing The Rising Sun.
Photo courtesy Roy Haviland

Graves Commission had already adopted 31st August 1921 as being the cut-off date for determining the official list of war dead.

The Cenotaph occupies an interesting legal position in the community. The land on which it is built and the adjacent car park is owned by Bream Sports Club, purchased from the Forestry Commission along with all the Sports Club ground in 1975.[3] However nobody actually 'owns' the Cenotaph itself. In March 1998, at the request of the British Legion, West Dean Council accepted responsibility for the upkeep of the Cenotaph but this responsibility is not based on ownership. It is the Parish Council recognising the community value of this great memorial that stands in our midst and supplementing the work tirelessly done by Melville Watts, Lloyd Jenkyns and others over the last 30 years.

If anyone can be said to 'own' the Cenotaph, it is the residents of the Ecclesiastical Parish of Bream. It was the Parish, as constituted at the time of the First World War, that built the monument by public subscription. The collections took place in Bream Village and all areas around – Clement's End, Little Drybrook, Brockhollands and Saundersgreen.[4]

Restoration work on the Cenotaph in 1970. *Photo courtesy Melville Watts*

1989 restoration of the 1939-45 lettering. *Photo courtesy Arthur Vaughan*

Over the years. the community has contributed magnificently to the upkeep of the Cenotaph – as they are doing now. Work was carried out on the monument in 1970, when the lettering was re-cut and the stones repointed. In 1982 the base was concreted and in 1989 the lettering on the north side, the Second World War names, was again re-cut. In 1994 further work was undertaken and the whole area made environmentally more attractive. In 1995 the north side lettering had again suffered damage and was replaced by a bronze casting of the names.(5)

Bream Cenotaph was honoured in 1995 when it was adopted as the home for the Forest of Dean Burma Star Association plaque. Bream was chosen because the view from Sun Tump reminded the Burma veterans of the hills, valleys and forests of Burma.

The Forest of Dean Branch of the Burma Star Association had been formed in August 1955 at the New Inn in Whitecroft.[(6)] Initially, it had over 130 members and it held meetings around the Forest in different pubs, the location usually determined, according to the Secretary, Arthur Vaughan, by the quality of the beer! The Association's standard was dedicated at Lydney Church on 20th March 1960, in what was possibly the largest parade Lydney has ever witnessed. Time passes, there are now only 28 members of the Association. In a couple of years' time the standard will finally be laid up for the last time.

The dedication service for the Burma Star plaque was at 7pm on Sunday, 2nd July 1995. The service was conducted by the Rev. Alistair Kendall. Bream Band and the Forest of Dean Male Voice Choir were in attendance. Some four hundred people were present.

The plaque is:

'In lasting memory of all our comrades who served in the Burma Campaign 1941-1945.'

It also bears the Kohima epigram adopted by the 2nd Division for their memorial in Kohima itself:

'When you go home tell them of us and say
For your tomorrow we gave our today.'

The original epigram had been written by the Greek poet Simonides in 480 B.C. to commemorate the three hundred Spartans who gave their lives at Thermopylae against the Persians.

Bream Cenotaph is a magnificent memorial in a magnificent setting. The challenge the community faces at the beginning of the millennium is every bit as great as the challenge of building it 80 years earlier. The very magnificence of its position makes it vulnerable to the weather.

John McCrae was a Canadian doctor who served in the Western Front. His poem *'In Flanders Fields'* inspired the use of the poppy as a symbol of remembrance. The third verse of the poem is a fitting reminder of the challenge we now face with the Cenotaph. The foe for us is a different one – the weather!

'Take up our quarrel with the foe!
To you from failing hands we throw
The torch; be yours to hold it high.
If ye break faith with us who die
We shall not sleep, though poppies grow
In Flanders fields.'

The Burma Star Dedication ceremony of July 1995.

Top: The Service of Dedication.

Left: The Plaque.
Photos courtesy Arthur Vaughan

Burma Star veterans at Bream Cenotaph, Remembrance Sunday, 11th November 2001. Standing, left to right: Fred Parry, Harry Tremblett, Cecil Hopkins, Derek Preece, Geoff Aldridge, Bert Cole and Mick Baldwin. In front: author Ian Hendy and Arthur Vaughan. *Photo courtesy Steve Cassidy*

The wreath layers on the same day. Left to right: Lloyd Jenkyns, Les Harper, Jane Watts and Melville Watts. *Photo courtesy Steve Cassidy*

The Burma Star veterans chose Bream Cenotaph for their plaque because the view from it reminded them of the Burmese hills, as demonstrated by this July 1995 photograph. *Photo courtesy Arthur Vaughan*

References

Extensive use has been made of G.R.O. Indexes of Births, Marriages and Deaths, Census Returns, the Commonwealth War Graves Commission Database, Soldiers Died in the Great War, the Glosters Database in the Regimental Museum and standard books on both World Wars.
Specific references that are not included in the text are listed below.

Introduction

1. Log Book, Bream Mixed School, Gloucester Records Office S57/1/3.
2. Second World War anecdote from Dave Rudge.

A Note on Money

1. National Statistics, The Economy Dataset.
2. A.L. Bowley, *Wages & Income in the U.K. Since 1860.*
3. B.L. Mitchell and H.G. Jones, *Second Abstract of British Historical Studies.*

Wenty's Sturty Bird

1. *Children's Games with Things, Marbles, Five Stones, Throwing and Catching,* Iona and Peter Opie.

1914 – *"Over by Christmas . . ."*

1. Kelly's Directory, 1914.
2. *The Mercury*, 18th September and 6th November.
3. *The Mercury*, 18th September and 6th November.
4. *The Mercury*, 18th December.
5. *The Mercury*, 27th November.

1915 – *"Only 4 Million Germans Left to Fight . . ."*

1. David Scott Daniel, *Cap of Honour*, 1951.
2. David Scott Daniel, Op. cit.
3. *The Mercury*, 18th June.
4. *The Mercury*, 27th August.
5. *The Mercury*, 3rd and 10th September. War Diary, 7th Glos.
6. *The Mercury*, 18th December 1914; 26th February, 5th March, 19th March and 17th December, 1915.

7. *The Mercury*, 27th August.
8. *The Mercury*, 24th September.
9. *The Mercury*, 22nd October.
10. *The Mercury*, 18th June, 12th November and 3rd December.
11. *The Mercury*, 30th July.

1916 – "*What Desolation in Homes War Makes . . .*"

1. *The Mercury*, 1st August, 1941.
2. War Diary, 7th Glosters.
3. *The Mercury*, 1st December 1916.
4. Battalion War Diary, March 1916.
5. Mixed School Log Book, 31st May and 1st June.
6. *The Mercury*, 19th May.
7. *The Mercury*, 14th July.
8. *The Mercury*, 22nd September.

1917 – "*Some Fitting Memorial to Those Local Lads Who Had Given Their Lives . . .*"

1. Quoted in Toni and Valmai Holt, *Battlefields of the First World War.*
2. Ron Hewlett.
3. *The Mercury*, 3rd August. Personal details on Willer from Margaret Aude, Eve Edmunds, Rona Followes and Ken Kear.
4. *The Mercury*, 31st August 1917.
5. *The Mercury*, 20th August 1915.
6. Biographical detail provided by Dave Rudge.
7. *The Mercury*, 19th October.
8. P.R.O. WO 97/5282.
9. Ron Hewlett.
10. My thanks to Jean Cook, Ken Hootton and Lynn Paddock for their help in contacting present day members of the Kilby family, and to Graham and Wayne Kilby for providing biographical details of the family. *The Mercury*, Refs. 11th August 1916, 18th May 1917 and 25th May 1917.
11. Sources on Angus Dodgshon include Wenty and Angus Cooper (who was named after Angus Dodgshon); P.R.O. WO 274/20102; *The Mercury*, 18th September, 6th November, 27th November, 18th December 1914; 29th October 1915 and 23rd November 1917.
12. Sources used: *The Mercury*, 29th November 1914, 5th February 1915, 18th June 1915 and 30th March 1917.
13. *The Mercury*, 18th May, 1st June and 8th June 1917.
14. *The Mercury*, 9th March 1917.
15. *The Mercury*, 13th July, 12th October and 16th November 1917.

16. *The Mercury*, 16th November 1917.
17. *The Mercury*, 27th July 1917.

1918 – "To the Last Man . . ."

1. 13th Glos War Diaries.
2. Mixed School Log Book, S 57/1/3.
3. *The Mercury*, 29th January 1915, 9th March 1917, 19th October 1917, 2nd November 1917 and 18th February 1918. Family details from Nancy Merry.
4. Family details from Margaret Hopkins. *London Gazette*, 18th June 1917. *The Mercury*, 5th April 1918. P.R.O. WO 363 T0290.
5. David Wilks.
6. Ron Hewlett.
7. *The Mercury*, 17th May 1918.
8. *The Mercury*, 17th May 1918.
9. *The Mercury*, 21st June 1918.
10. 12th Battalion Glosters War Diary. *The Mercury*, 18th December 1914, 22nd January 1915, 5th October 1917 and 19th July 1918.
11. Family details from Joan Edey. Postcard information from Gwen Warren. Service Papers P.R.O. WO 363/F775.
12. *The Mercury*, Granny Kent Obituary held by Rex Morgan.
13. Basil James, Ron Hewlett and Jack Wildin.
14. Battalion War Diary P.R.O. WO 95/2577.
15. I am grateful to Rex Morgan for giving me access to Joe's letters and poems. Other references used: *The Mercury*, 21st August, 4th December 1914, 4th February, 26th February 1915 and 17th May 1918.
16. Assistance in tracing Ernest Lucas was given by Donald Lucas.
17. Infants and Mixed School Log Books.
18. *The Mercury*, 15th March and 5th April 1918.
19. *The Mercury*, 31st May 1918.
20. *The Mercury*, 22nd February 1918.
21. *The Mercury*, 16th August 1918.
22. *The Mercury*, 25th July 1918.
23. Infants School and Mixed School Log Books, 31st July 1919.
24. *The Mercury*, 15th October 1920.
25. *The Mercury*, 29th October 1920.
26. *The Mercury*, 12th November 1920.
27. *The Mercury*, 11th November 1921.
28. *The Mercury*, 9th September and 16th September 1921.

1919 - 39 – *"A Truce For Twenty Years"*

1. Quoted in Tom Harrison, *Living through the Blitz*, 1976.

The Phoney War – September 1939 - May 1940

1. Kelly's Directory, 1939.
2. G.R.O. S57/2/7.
3. *The Mercury*, 2nd February 1940.
4. *The Mercury*, 20th October 1939.
5. *The Mercury*, 15th February 1940.

Britain Alone – 1940-41

1. Translation provided by Satu Pohjola.
2. Biographical details and family correspondence on Morris provided by Clive Bath and Molly Bath.
3. Ron Hewlett.
4. Len Deighton, *Fighter.*
5. Hylton Miles.
6. Family details provided by Joan Lee.
7. Hylton Miles.
8. Ron Hewlett.
9. John Ray, *The Night Blitz*, 1996.
10. Family details from Lloyd Jenkyns.
11. Charles Hocking, *Disasters at Sea.*
12. Biographical details from Ken Kear, Hylton Miles and Des Cassidy.
13. *The Mercury*, 27th December 1940.
14. Personal details on Bill from Elsie Drew, Pat Lewis, Hylton Miles and John Palmer. R.A.F. sources: W.R. Chorley, *R.A.F. Bomber Command Losses, 1941*, and Ministry of Defence AHB3 (R.A.F.) 8/27.
15. Ludovic Kennedy, *Pursuit, The Sinking of the Bismark*, 1975.
16. Personal details on Bob from Harry Carpenter, Ann Sargent and Geoff Watkins.
17. *The Mercury*, 30th May and 6th June. Personal details on Doug from Trevor and Joan Davies.
18. Ron Hewlett.
19. Personal details on Tommy from Ella Dawe and Geoff Wildin.
20. *The Mercury*, 12th July 1940.
21. *The Mercury*, 26th July 1940.
22. Evacuee details from Infants and Mixed School Log Books. Personal accounts from Len and Mick Olley, Sylvia Parry and Dave Rudge.
23. *The Mercury*, 27th September 1940.
24. *Songs, Ballads and Parodies of the Great War*, EMI 1978.
25. *The Mercury*, 11th October 1940.
26. *The Mercury*, 11th October 1940.
27. *The Mercury*, 14th February 1941.

28. *The Mercury*, 7th February 1941.
29. *The Mercury*, 7th March and 27th June 1941.
30. *The Mercury*, 26th September 1941.
31. *The Mercury*, 30th August 1940.
32. Home Guard details, courtesy of Pat Lewis.
33. *The Mercury*, 22nd August 1941.
34. *The Mercury*, 3rd May 1940.
35. *The Mercury*, 16th August 1940.
36. *The Mercury*, 18th July 1941.

The Turn of the Tide – 1942 - 43

1. Personal details on Percy from Alice Davies and Ron Hewlett.
2. Personal details on Reg from Hylton Miles, Hazel Sterry and Betty Wooles.
3. Personal details from Ronald and Ann Wilks. Also *Lydney Observer*, 8th January and 15th January 1943.
4. Ken Delve and Peter Jacobs, *Bomber Command in World War Two.*
5. Personal details on John from Reg Overington. R.A.F. sources: W.R. Chorley, *Bomber Command Losses, 1943.*
6. Personal and Service details on Walter from Geoff Wildin. Charles Cruikshank, *The German Occupation of the Channel Islands.* J.C. Salvary, *Diary of the German Occupation of Guernsey.* Peter King, *The Channel Islands War.*
7. *The Mercury*, 14th January 1944.
8. *The Mercury*, 27th February and 27th March 1942.
9. *Lydney Observer*, 12th February and 4th June 1943.
10. *The Mercury*, 7th August and 11th September 1942.
11. *Lydney Observer*, 23rd July 1943.
12. *Lydney Observer*, 16th April 1943.
13. *Lydney Observer*, 14th May 1943.
14. *The Mercury*, 16th October 1942. Lydney Observer, 6th January 1943.
15. *The Mercury*, 27th February 1942. Lydney Observer, 9th April 1943.
16. *Lydney Observer*, 16th April, 4th June and 22nd October 1943.
17. *Lydney Observer*, 13th August 1943.
18. *The Mercury*, 13th November 1942. Lydney Observer, 15th October 1943.
19. *The Mercury*, 13th November 1942.
20. *The Mercury*, 27th November 1942. Lydney Observer, 3rd December 1943.

The Collapse of the Axis – 1944-45

1. Bryan Perrett and Ian Hogg, *Encyclopaedia of the Second World War*.
2. Personal details on Cyril from Gloria Marshall. A Personal Account of the Siege from (Sgt) John Willcock. David Rooney, *Burma Victory* (1992). Arthur Swinson, *Kohima* (1966). Arthur Campbell, *The Siege, a Story of Kohima* (1956).

3. Personal and Service details on Cyril from David Morgan. Also Gavin Dawes, *Prisoners of the Japanese.* Graham N. Thompson, *Prisoners of the Rising Sun* (*Family Tree Magazine*, November 1999).
4. Chester Wilmot, *The Struggle for Europe* (1952).
5. Ron Hewlett.
6. Personal and Service details on Stan provided by Elsie Drew.
7. Personal and Service details on Des from Betty Camm and Iris Hoare. David Scott Daniel, *Cap of Honour* (1951).
8. Personal and Service details on Nobby from Iris Burford and John Baker.
9. Gerald Vaughan.
10. Personal, Family and Service details on Verdun provided by Terry James, Margaret Joyce, Dolly Milner and Edna Pring. Polish captivity details from William Prankard. David Scott Daniel, *Cap of Honour* (1951). *The Back Badge*, Winter 1978.
11. Details on Tommy from Dennis Challenger.
12. Details on Den from Ron Hewlett, Ken Kear and Hylton Miles.
13. Personal details on Dora from George and Nancy Hoare, and Maureen Williams. Obituary, *Lydney Observer*, 7th June 1946.
14. *The Mercury*, 26th May and 16th June 1944.
15. *The Mercury*, 7th January 1945.
16. *The Mercury*, 30th June 1944.
17. *The Mercury*, 28th July, 3rd November and 8th December 1944.
18. *The Mercury*, 3rd November 1944.
19. Infants School Log Books.
20. *The Mercury*, 20th July 1945.
21. *The Mercury*, 29th December 1944.
22. *The Mercury*, 11th May 1945.
23. Infants School Log Books.
24. *The Mercury*, 19th October 1945.
25. *The Mercury*, 14th June 1946.

Bream Cenotaph

1. Sources used on the War Memorials include: Alan Borg, *War Memorials* (1991), Colin McIntyre, *Monuments of War: How to Read a War Memorial.*
2. I am indebted to Harold (i.e. John!) and Joan Davies for this information.
3. Michael Walby.
4. Parish Map provided by Reverend Alistair Kendall.
5. Treasurer's Accounts kept by Lloyd Jenkins.
6. Details of the Forest of Dean Burma Star Association provided by Arthur Vaughan.

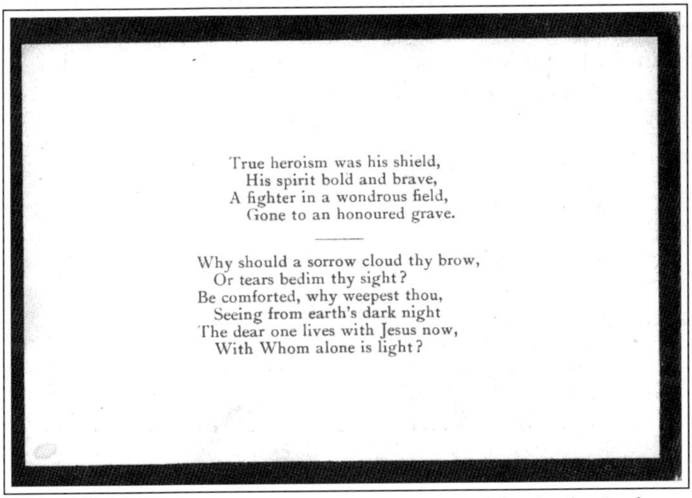

True heroism was his shield,
His spirit bold and brave,
A fighter in a wondrous field,
Gone to an honoured grave.

Why should a sorrow cloud thy brow,
Or tears bedim thy sight?
Be comforted, why weepest thou,
Seeing from earth's dark night
The dear one lives with Jesus now,
With Whom alone is light?

Front and reverse of the 'In Memoriam' card for Leslie Lodge.
Courtesy Nancy Merry